Nanotechnology, Law and Legal Services

Nanotechnology, Law and Legal Services

Shailendra Sengar

ANMOL PUBLICATIONS PVT. LTD.
NEW DELHI-110 002 (INDIA)

ANMOL PUBLICATIONS PVT. LTD.
Regd. Office: 4360/4, Ansari Road, Daryaganj,
New Delhi-110002 (India)
Tel.: 23278000, 23261597, 23286875, 23255577
Fax: 91-11-23280289
Email: anmolpub@gmail.com
Visit us at: www.anmolpublications.com

Branch Office: No. 1015, Ist Main Road, BSK IIIrd Stage
IIIrd Phase, IIIrd Block, Bengaluru-560 085 (India)
Tel.: 080-41723429 • Fax: 080-26723604
Email: anmolpublicationsbangalore@gmail.com

Nanotechnology, Law and Legal Services

First Published, 2008

Reprint, 2012

ISBN 978-81-261-3436-6

PRINTED IN INDIA

Contents

Preface

This Encyclopaedic Series of Career and Vocational Guidance Covers careers, describing jobs, places of employment, working conditions, qualifications, education and training. This has been developed to aid career and counsellors and job placement coordinators in planning activities for students and potential students about career vocational education programs. It contains a career and vocational guidance model which clarifies what career and vocational guidance services mean and what planners should include in their activities. The model shows an interrelatedness among major thrusts with career development being a central theme of vocational guidance. All of the thrusts depicted in the model represent essential services that should be available to students in career and vocational education programs. The encyclopaedia deals with various sections which correspond to the major thrusts identified in the model: (1) career development; (2) recruitment; (3) guidance and counselling; (4) placement; (5) follow-up and follow-through; and (6) consultation. Each section contains an outline of the perceived needs of the topic, an overview of the topic relative to career and vocational guidance, and some suggested activities and strategies that may be useful in meeting student needs.

It also contains current information about occupations, employment trends, and job searching. Materials on career decision-making, internships, and salaries are also located in each volume. This Series of Encyclopaedia helps students assess their potential for various careers.

Shailendra Sengar

1

Career in Law and Legal Sector

Law

Legal profession is one of the growing and lucrative professions all over the world. It is one of the most adventurous as well as exciting career. Lawyers are held in high esteem in our society and there remains the faith that when all else fails, one can still take recourse to the legal system. In our daily life in one way or other we may come across situations where legal advice is required. In such situation we need the advice of a legal professionals who analyse and interpret law properly.

Lawyers counsel clients about their legal rights and suggest course of action in a particular matter. They also advise clients and draw up legal documents, such as wills and contracts and conduct negotiations on legal matters and may represent clients in court and tribunal proceedings.

Academic excellence is not the main thing behind the success in this profession. Professional competence acquired through experience and practice with efficient lawyers is the main yardstick of success.

There are various areas of specialisation in this field which one can choose from like corporation law, civil law, criminal law, international law, labour law, patent law, tax law and so on. There are number of institutions and universities conducting courses in law. The most premier among law universities in India is National Law School of India University

(NLSIU), Bangalore. Admission to NLSIU is through an All India Admission Test held usually on first Sunday in May. University College Bangalore, Government College Mumbai, ILS law college and Symbiosis college, Pune and Faculty of Law of Delhi University, Aligarh Muslim University etc. are some of the other leading legal Institutions in India.

Eligibility and Course Areas

Educational: Minimum qualification to go for a course in law (LLB) is 10+2. Course duration is of 5 years and is sponsored by the Bar Council Of India.

P.G. Programmes in law (LLM) are of 2 years duration & eligibility is LLB degree.

In addition to LLB degree, many universities and institutes also offer other diploma courses in several disciplines of law, including Administrative Law, International Law, Labour Laws, Tax Laws and Corporate Laws.

In most of the institutes/universities, selection is on the basis of an entrance test. The entrance examination is objective type in nature. The questions invariably cover areas like Reasoning, General Awareness, Numerical Aptitude, Legal Aptitude and Preliminary Political Science.

Teaching is not restricted to the conventional classroom methods alone, it include case studies, mock court etc. to make the study more practical and interesting.

There are various branches of specialisation in law like civil law, criminal law, corporate law, property law, income tax law, marine law, public international law, family law, labour law, press law, excise law, constitutional law, administration law, sale of goods law, trade mark, copyright and patent law etc. one can choose from.

Personal Attributes: Success in this profession is not based on marks one obtain in the exam, but it is very much related to personal attributes. One must have good oral and written communication skills, logical reasoning, power of concentration, patience, good memory to relate and quote past cases to prove your point, ability to argue and discuss matters with a variety of people, self confidence, courage to deal with threats especially in criminal cases etc. He should have up-to-date information on any changes in law. A good library and a fair amount of reading is also important.

Job Prospects and Career Options

Job Prospects: They can find employment in various courts of law, in government service both Central and State government, as teachers, as legal advisors in various companies, business houses, organisations etc. They also do private practice as legal advisors, advocates, solicitors etc.

Career Options: One who wishes to become an advocate and practice law as a profession in India, should enroll with the local State Central Bar council. An application for admission as an advocate shall be made in the prescribed form to the State Bar Council within whose jurisdiction the applicant proposes to practice. The Bar Council of India and the State Bar Councils forms a self contained code of law for legal professionals practicing in India.

Lawyers can find employment in a variety of positions in the Government Service as well as other fields. Lawyer can also work as legal counsel and legal advisers for firms, organizations and families. They can work as trustees of various trusts, as teachers, law reporters, company secretary and so on. Opportunities exist in the defense services too. Additional qualification along with law degree offers scope for a wide range of employment opportunities.

Government Services: Lawyers are appointed in central services (Indian legal service) through UPSC on grounds of their experience in the field. Post covered under Indian legal service are Law officers, Assistant Advisors, Dy. Legal Advisors and Legal Advisors.

State legal service appoints lawyers through SPSC exams in senior position in State police/Revenue/Judicial Departments. The basic qualification is degree in law, besides a degree in Arts / Science / Commerce / Agriculture. Average age limit is between 21 to 30 years.

In the judiciary, the posts are that of the magistrate, district and sessions judge, munsifs (sub-magistrate), public prosecutor, solicitors, attorney general, advocate general, notary and oath commissioner. There are also legal secretaries appointed in assemblies, law inspectors, legal officers in banks, judicial members of income tax, sales tax and excise departments, government advocates and staff in the registrar of companies' office. The examination for these posts take place periodically & is advertised as and when the vacancies occur.

Magistrate and Sub-Magistrate: A candidate who qualifies in the Law Service Commission or State Public Service Commission is eligible for appointment as Magistrate and Sub-Magistrate (munsif). While magistrate presides over criminal court, munsif delivers judgment on civil cases. Promotion may take a person higher up to sub-judge, District and Sessions Courts Judge and further (depending upon seniority and vacancy) to appointments in High Courts and the Supreme Court.

Public Defenders: At the state, federal and local level, public defenders serve as court-appointed attorneys for those who cannot afford an attorney.

Public Prosecutor: Govt. Pleaders represent the State in sessions courts and lends advice to officers in districts in all legal matters when called for.

Solicitor: Advices the government on legal matters and submit statement of important cases to Attorney General of India for his opinion or appearance in the Supreme court as necessary. A Solicitor General makes sure that the Attorney General performs his duties properly.

Attorney General: Attorney general appointed by the president advices the government of India on such legal matters as may be assigned to him by the president. He has the right of audience in all courts of India and participate in the proceedings of Parliament without the right to vote.

Advocate General: Similar to the post of Attorney General at centre, each state has an Advocate General appointed by State Governors. These offices generally prosecute cases that have statewide significance and work with local District Attorney's offices in the prosecution of cases. One of the best ways to gain trial experience quickly after graduation is to work at a state attorney's office.

District Attorney: Assistant State Attorneys (also called district attorneys) are primarily responsible for criminal matters and prosecution. The District Attorney is usually an elected official. Depending upon the size of the District Attorney's office, they may have multiple units that focus upon different types of crime such as domestic violence, homicide, appellate work or sex crimes. If you work with a smaller District Attorney's office, you may have the option of handling diverse caseloads.

Notary, Oath Commissioner: A notary is a legal practitioner of at least ten years experience in the profession. He is appointed on application to Central, State government to draw, verify, authenticate, certify and attest the execution of any deed. Oath commissioner is another established authority similar to the notary.

Legal Advisers: Public sector undertakings and state and central government organisations also employ lawyers as advisers.

Other Fields: Lawyer can also work as legal counsel and legal advisers for firms, organizations and families. They can work as trustees of various trusts, as teachers, law reporters, company secretary and so on. Opportunities exist in the defense services too. Additional qualification along with law degree offers scope for a wide range of employment opportunities.

Defence Service: They can work in the legal branches of defence departments such as the Indian army, Navy and Air force. They conduct courts of enquiry and court martials of offending service personnel governed by the particular legislation acts.

Legal Advisors: They work in legal departments of firms or for private clients. Legal advisors counsel their clients as to their legal rights and obligations and suggest the appropriate course of action in business and personal matters. They also check deeds, issue writs, collect information for affidavits and draft legal documents.

Advocate: An advocate practises in the court of law. He represents a party (individual, group of individuals or a company) in a court of law. They present the facts of the case to the judge, cross examine witnesses and finally sum up the reasons as to why the court should decide in their clients' favour.

Solicitor: Solicitor is a kind of advisor to clients on a variety of matters that involves law. Solicitors explain the cases of his clients to the advocates. Solicitors need never appear in court directly. Solicitor's firms undertake cases related to their area of work.

Teacher: Those with a good aptitude and taste for teaching, can go for LLM and take up job in any University or Institute offering law courses to students.

Writer: If he has a flair for writing may act as an author and produce journals and legal commentaries and bring out law reports. They can write articles for magazines, news papers etc.

Company Secretary: A law graduate and/or a member of the Institute of Company Secretaries can be employed as a company secretary in a firm.

Remuneration

Salaries of experienced attorneys vary widely according to the type, size and location of their employer. Initially an articled clerk or apprentice earns a stipend of Rs 2,200/- This will increase substantially over years of experience. In the government judicial service salaries range from Rs 2,000-4,000. A sub judge earns Rs 5,500 along with the usual allowances and perquisites applicable to government employees. Most salaried lawyers are provided health and life insurance and contributions are made on their behalf to retirement plans. The private legal practitioner earns comparatively handsome amounts.

Industry Overview: Law is the way society regulates its behaviour. The intent of law is to create rules of conduct that are widely understood and respected throughout the jurisdiction in which they were written and to create normalized processes for adjudicating disputes. Because law is a fairly technical profession not necessarily easily comprehended by the untrained, individuals and companies hire professionals-lawyers-to help them understand it and follow the procedures it defines.

Most of what lawyers do is paperwork. They research legal precedents, spending hours or months in law libraries or with online databases. They prepare contracts, briefs and other documents, assembling boilerplate paragraphs or writing text from scratch. They plan and conduct depositions, which in complicated cases can generate thousands of pages of testimony, all of which has to be read, analyzed and refined into usable information. Sometimes, especially if they are litigation specialists, lawyers actually argue cases before judges or juries.

To practice law, you must pass the bar exam in the state in which you want to practice. Almost all lawyers earn their JD degree after three years of law school and then take the bar exam in the state in which they wish to practice. In general, the better the law school you graduate from and the higher your class rank, the better your job prospects once you graduate.

Many lawyers work for big corporate law firms or as in-house attorneys in corporations, while others hang out their own shingles and represent individuals or small companies in divorces, bankruptcies, estate planning and the like. Other lawyers work for the government, in various agencies or as district attorneys or public defenders. Still others work for advocacy groups such as the ACLU or the NAACP. In addition, some lawyers-who've gone to law school and passed the bar exam-do not practice law, but choose to work instead in business, banking, academia, or politics. Then there are paralegals, high-caliber support people who do everything from word processing to legal research. Paralegals sometimes decide that they enjoy the field so much that they end up going to law school themselves.

Career Overview: Although television dramas might have you believe that you only need good looks and a fashionable wardrobe to succeed as a lawyer, in reality, completing three years of law school and passing a tough bar exam barely prepare you for the practice of law. Law is the way society regulates its behaviour. The intent of law is to create rules of conduct that are widely understood and respected and to create normalized processes for adjudicating disputes. Because law is considered a technical profession not easily comprehended by the untrained, individuals and companies hire professionals-lawyers-to help them comprehend it and conduct the procedures it defines.

The law is an integral part of nearly every area of our lives-from the environment, trade and commerce and civil rights to national security, cyberspace and entertainment and sports.

People who enter this demanding and all-encompassing profession do so for many reasons, not the least of which is the desire to work in an intellectually rigorous field that can have a significant impact on the world around them. Of course, in most legal careers, the pay is great, too, but most lawyers will tell you the money isn't enough to sustain one's commitment to a job that can be incredibly challenging, with long hours and high stress. No, if you want to be a lawyer, you'd better love the law.

Lawyers can work for law firms (sometimes very big law firms), in government agencies, in legal departments of private corporations, for corporate clients (*e.g.* advising corporate management on the acquisition of a competitor), for individuals (*e.g.* for the defendant in an assault-and-battery case), or they can hang out their own shingles.

What You'll Do?: In broad terms, lawyers apply their interpretation of the law (the codified rules of their society) to advise their employer or client on completing transactions in compliance with the law or resolving disputes based on current understanding of the law.

Your Legal Career: The legal field is an exciting place to be. Salaries are up, the hiring outlook is good and there are many options to consider, from attorney to paralegal to legal secretary and more.

Whether you're a student or career changer looking to get into the field or a legal professional aiming to maximize your career and achieve work/life balance, these articles will give you the boost and insight you need:

- o Law Hiring Outlook 2007
- o Law Interview Do's and Don'ts
- o Ten Resume Laws
- o Legal Salary Trends
- o Temp in the Legal Industry
- o Law Partnership

Law Hiring Outlook 2007

If you're a lawyer or third-year law student, prospects for finding a new job in 2007 are good. If you're a paralegal, legal secretary or other skilled support staff, your prospects are great.

That's the consensus of observers as business transactions drive the demand for competitively priced legal services, both at private firms and in corporate departments.

"The prospects for 2007 hiring in the legal arena are high," says Steven Levine, vice president of national permanent operations for staffing firm Hudson in New York. "There's a great demand for candidates with experience in mergers and acquisitions and corporate litigation, for both law firms and corporate legal departments."

Legal professionals are generally bullish on employment prospects in their industry, according to the Hudson Employment Index. Some 31.4 percent of respondents to the December 2006 survey said they expected their employers to be hiring in coming months; only 11.6 percent anticipated layoffs—just more than 50 percent expected no change. Still, nearly 20 percent of respondents said they were worried about losing their jobs.

Attorneys in Demand

With corporate compliance requirements increasing and business transactions evolving with global markets, demand for skilled attorneys and the best and brightest law students is healthy.

"In the past, we could largely restrict our hiring to a couple of months in the fall of each year," says Racquel Keller, manager of professional development and recruitment at Sterne, Kessler, Goldstein & Fox PLLC in Washington, DC. "Now our recruiting efforts have been driven to more of a year-round schedule."

Although many large firms hire the upcoming class of law graduates almost a year ahead, many other legal employers are still recruiting for this year. "Small and medium-sized firms and government agencies and public-interest organizations hire when they need somebody," says Randi Friedman, director of career services at Northeastern University's School of Law in Boston.

Private firms are also hiring more experienced lawyers to quickly build practice areas as the business environment shifts. "Over the last couple years, there has been a large shift toward lateral hiring" of experienced attorneys from other employers, says Keller.

"The corporate law world is such that certain kinds of transactions become hot and others slow down," says Scott Berson, a partner and chair of the recruitment committee at Chadbourne & Parke LLC in New York. "Right now, private equity and hedge funds are hot."

Legal recruiters could almost take their orders by reading business news headlines. "Compliance and options backdating are hot areas," Levine says. Still, "corporate legal departments are looking for people with rounded experience."

Although certain attorney specialties will be in high demand for years, overall growth in attorney jobs will be constrained by thrift. "More and more employers realize that they can get some legal work done less expensively with paralegals than with attorneys," says Rafia Aleem, a paralegal manager and member of the University of California Irvine Extension's paralegal advisory committee.

Flush Times for Paralegals and Legal Secretaries

Indeed, skilled professionals who support lawyers are highly sought after, as both private firms and in-house departments respond to competitive pressures to keep legal fees contained.

"There's a tremendous need for paralegal and legal secretaries, much more of a demand on the corporate side than in the past," says Levine.

Professionals without law degrees form a major employment sector in the legal industry. Paralegals and legal assistants held about 224,000 jobs in 2004, according to the 2006-07 edition of the Bureau of Labour Statistics' Occupational Outlook Handbook. More importantly, growth of these jobs is projected to be much faster than overall employment growth through 2014.

Recent job openings include legal secretaries specializing in each of the following specialties: corporate, litigation, intellectual property and patent prosecution. Paralegal openings recently listed include specialists in litigation, real estate, trademark, bankruptcy and loan administration.

Law Interview Dos and Don'ts

In some ways, the interview process for an attorney, paralegal, legal assistant or other law-based position is similar to interviewing for other jobs. You need to be well-prepared, solid with your responses and smart in your follow-up. But legal interviews carry their own unique nuances and expectations as well. Here are a few key dos and don'ts to keep in mind.

Interview Dos:

- Dress Conservatively: If you're a man, go with a navy or black suit, a white shirt and "a simple striped or solid tie that is bold but not loud," says Ed Shioyazono, senior director of Direct-Hire Staffing for legal placement firm Hudson. Woman will want to accompany a conservative skirt suit with highly polished shoes and "very minimal jewelry," he adds. "Law firms may have five-day business casual, but the 'uniform' is expected for all interviews."
- Go in with a Game Plan: Legal interviewers use a variety of interviewing styles, but the most common is a conversational, nondirected approach, says Bridget Kenadjian, assistant director of career services at the William Mitchell College of Law in St. Paul, Minnesota.

Knowing this may put you at ease, but the downside is you risk not being able to get your key messages across. So you need to have a game plan "regarding what messages you want the employer to receive," says Kenadjian. "Look for ways to slip in the content that you know [the interviewers] need to have in order to hire you—even if they don't ask the right questions."

- *Be Ready to Address Weaknesses Without Getting Defensive:* The legal job market is extremely competitive, so you need to carefully consider—from the employer's perspective—what your potential weaknesses might be and then be ready "to tactfully and gracefully overcome those objections in an interview," says Kenadjian.

  "Such questions or statements as 'Why weren't your grades better?' or 'You don't seem to have much experience in X' may need to be addressed without [your] becoming defensive," Kenadjian adds. "Try to predict what they might be and prepare in advance effective, non-argumentative responses to them."

  Interview Don'ts:

- *Be Late—Ever:* In fact, arriving for your interview at least 15 minutes early "makes a stronger impression than you might think," advises Shioyazono.

  "In a law firm and the practice of law generally, everything revolves around billable time—time billed to the client for work completed—and deadlines," Shioyazono explains. "Arriving early communicates that you do not, or likely will not if you are a junior (associate), have a problem underestimating the importance of scheduling, managing and billing time."

- *Criticize Others:* In particular, "don't bad-mouth your former employer," says Maureen Brady, division manager for Hudson Legal's Permanent Placement Division. Doing so will make your interviewer wonder what you'll say about him if you're hired.

  Perhaps more importantly, the person you're bad-mouthing might be a close friend of the person you're interviewing with. Remember: The legal world is surprisingly small.

- *Forget to Thank Your Interviewers:* Send a thank-you note via snail mail within 24 hours of your interview, says Deborah Schneider, author of Should You Really Be a Lawyer?

  "I can't believe how many people overlook this basic interview etiquette," says Schneider. "The note doesn't have to be long—several sentences on a note card are fine—but sending one is a must. I know plenty of candidates who did not get a second interview, because they failed to send a thank-you note after the first."

Ten Resume Laws

When you're looking for a new legal position, the quality of your resume can make or break your job search. Follow these 10 tips from recruiters to produce a winning legal resume that will help you stand out.

Make a Good First Impression

When you receive up to 700 resumes a week, as does Hudson Legal placement director Penny Burke, you must make quick decisions on which ones warrant further review. "First impressions are paramount," including the resume's format, design, font and organization, Burke says.

Stacey Beck, recruitment director for the staffing firm's Philadelphia office, agrees that first impressions count, adding that experienced job seekers can be eliminated if they show poor judgment on their resumes. "I review an average of 25 resumes daily and I can tell within several seconds whether I'm going to interview the candidate or not," she explains.

Keep it Simple: Candidates often use flowery fonts, boldface, boxes, lines, colours and other embellishments to stand out. This is not recommended.

"Keep it simple and professional—nothing flashy, no pictures or graphics and no unusual fonts," advises Maureen D. Brady, division manager of Hudson Legal's Permanent Placement Division within Legal Solutions in Manhattan. "Maybe those strategies work in the marketing or advertising arena, but not in the legal field."

Be Concise

Brevity is essential. Brady, who specializes in placing paralegals, law firm administrators and clerks, says those candidates should keep their resumes to one page.

Ed Shioyazono, senior director of Direct-Hire Staffing in Hudson Legal's Los Angeles office, specializes in placing attorneys. He says a two-page resume is OK if you can't make your information fit on one page.

Use an Addendum

Burke suggests using a resume addendum if you have extensive experience. "Add an addendum containing trial highlights, business deal accomplishments or other noteworthy achievements," she says. "It looks more professional to attach an addendum than to have a three-page resume."

Eliminate Errors: "The legal field is all about accuracy," says Shioyazono. One type can be enough to disqualify an otherwise strong applicant, he adds.

Burke says spelling and grammatical errors will ruin a paralegal's chances. She advises job seekers not to rely on spell-check, as this method won't catch blunders like cover letters that begin "Deer Ms. Burke."

Highlight Your Accomplishments

Candidates need to show how they meet employer's specific needs. "People tend to be too general on their resumes," Shioyazono says. "If you're a real estate attorney, don't just say, 'Involved in purchase and sale.' Instead, highlight the depth and breadth of your skill set and achievements. It's better to say, 'Documented all due diligence and closed transactions for $400-thousand industrial real estate deal.'"

Be Selective

Your resume is not your life story. "I don't care that you play chess or have a family," says Beck.

Shioyazono advises legal professionals not to go back more than 10 years on their resumes. "It's unnecessary to go into detail about experience that has nothing to do with your career goal," he says.

However, don't be too hasty in omitting all pre-legal experience. "For example, if you were a real estate paralegal or worked in a bank before you went to law school, include this information," Shioyazono says. "I would interview someone with this background for a real estate attorney job before someone who's fresh out of law school without this relevant experience."

Strategically Organize Your Resume

"If you're a new graduate, don't place your experience first," advises Burke. "Lead with your strongest selling points." You can create multiple resume versions if you have more than one career goal or specialty area. "If you have two different types of positions in mind, don't try to pack everything into a single document," Burke says.

Emphasize Stability

If you've been on a steady career track, highlight this continuity in your resume. Shioyazono says the absence of job-hopping is even more important than being a perfect match to an employer's list of desired qualifications. "Having job stability says 'I'm not a flight risk,'" he explains.

If your employer changed names because of a merger or acquisition, make it clear you didn't change jobs. According to Beck, "lots of people appear to be job-hopping when in fact they're not."

Be Honest

"Don't exaggerate your position title or responsibilities," cautions Burke. "I've interviewed people who overstated their experience, making claims like, 'Handled 10-plus jury trials,' when in fact they just observed these proceedings." The same goes for lying about degrees or fabricating employment dates to eliminate gaps. Says Brady, "there's a good chance you'll get caught, if not during the interview process, then after placement."

Legal Salary Trends

Will 2007 be a good year for young lawyers and paralegals? Most indications suggest it will be, whether legal professionals choose to go for the gold or trade a chunk of change for better work/life balance.

Lawyers' Salaries Are Moving Up Again

The nation's largest, most prestigious law firms tend to set starting associates' salaries nearly in tandem. In January 2007, talk of a substantial hike in junior associates' pay was echoing in conference rooms in New York, Washington, Chicago and Los Angeles.

"People get their reviews in December and the unhappy ones start calling around," says Ed Shioyazono, senior director of direct-hire staffing at Hudson Legal. "There will be a lot of associates moving to firms that pay better."

"Our starting salary for associates beginning September 2007 is $135,000," says Racquel Keller, professional development and recruitment manager at Sterne, Kessler, Goldstein & Fox PLLC in Washington DC. "But we're not in a lockstep programme. It's driven by what the market and clients will bear."

Salaries are higher in the New York area, but so are expenses. "Our starting salary is $160,000," says Scott Berson, a partner and chair of the recruitment committee at Chadbourne & Parke LLC in New York.

A number of other New York firms joined the January move to a $160,000 starting salary, according to a report in the New York Law Journal. "The first-year salary at Simpson Thacher has just gone to $160,000 from $145,000," says Shioyazono.

In lesser cities, the numbers are slightly less impressive. "Now new grads at big firms in Boston start at about $135,000," says Randi Friedman, director of career services at Northeastern University's School of Law in Boston. "But there's a huge disparity between public and private sectors. Public defenders and district attorneys start at around $35,000 in Massachusetts."

Even the most junior associate can potentially earn extra. "The bonus pool for associates is determined by a number of factors, like meeting billable-hours requirements," says Berson. "There are different levels of bonuses. For people who go above and beyond the call, or get into business development, recruiting or writing articles, there may be a super-bonus."

Lateral Moves May Mean Lower Pay

More lawyers are deciding relatively early on that partnership is not their primary goal.

"The law-firm life is becoming less and less desirable for a lot of people after just two or three or four years," says Shioyazono. Many lawyers who are still quite young decide that if they want a life outside the firm, they must leave it.

"I've helped people go from earning $150,000 a year in a firm to $85,000 or $90,000 doing compliance in-house," says Shioyazono. "Lawyers making a lateral move typically go from a larger firm to a smaller one, where salaries are generally lower but all over the place, because the hours requirements vary." Another common lateral move for attorneys is from a private firm to a corporate legal department. "At a large, publicly traded company with an in-house department, a lawyer with four to five years of experience would typically go in at $160,000, a pay cut from a big firm," says Shioyazono.

Paralegals Earn a Broad Range

Paralegal pay varies widely, from a working-class paycheck to an impressive salary that enables an affluent lifestyle.

"Entry-level paralegal salaries ranged from $22,000 to $30,000" in 2005, the latest year for which the Orange County Paralegal Association has data, says Rafia Aleem, a paralegal manager and a member of the University of California Irvine Extension's paralegal advisory committee. "Paralegals with three to five years' experience earn from $40,000 to $60,000."

Still, paralegals as a whole haven't made a lot of progress beyond inflation. "They're not really benefiting from increased demand" for their services, says Shioyazono.

But the cream of the crop of paralegals—the most-qualified individuals working at top firms in the biggest cities—can earn more than a basic middle-class salary. "For the best senior paralegal—not even a manager—we saw the salary hit $100,000 in late 2006," says Shioyazono.

Curiously, you wouldn't know times are good for legal professionals by asking them. In a December 2006 Hudson Employment Index survey, 51.2 percent of respondents described their own personal finances as fair to poor. Only 40.3 percent said their financial status was improving.

Law Partnership

Dreaming of the day you'll be offered partnership at your law firm? There's more to consider than the big money partners make, warns Marc Zamsky, senior vice president of Hudson Legal and a legal recruitment specialist in Hudson's Philadelphia office.

"Partnership can take on many forms and there are many considerations that go into whether you should become a partner, where you want to go into partnership and what partnership holds for your career," says Zamsky.

The Track Layout

The typical partnership track lasts between seven and 10 years, beginning with the summer associate position. Next, you'll go from first-year associate to senior associate and finally on to partner.

How many lawyers make the cut? About 121,500 attorneys worked at the National Law Journal (NLJ) Top 250 firms at the end of 2006, says NLJ spokesman Lee Feldman. "Within that group, there were roughly 37,500 equity partners and 84,000 associates and nonequity partners," he says. "So, over time, roughly 30 percent have eventually made partner for this group. But that doesn't mean that on any given year, 30 percent of associates are going to make partner."

Zamsky estimates that half of associates hired by small firms eventually become partners. Their average salary might be $80,000 or $90,000. Partners at the NJL Top 250 firms earned a median income of $1.5 million in 2005, Feldman says.

How to Make Partner

To stay on the partnership track, make yourself valuable and likeable. Bill no fewer than 2,000 hours a year, produce well-written, accurately researched work and hone your instincts in your niche practice area, says Andrew Jewel, vice president of national operations for Hudson Legal. A dash of political savvy and the ability to get along with everyone from paralegals to partners also help.

The most important item on your to-do list is business development. Hard-working, legally astute attorneys who bill lots of hours are easy to come by, Jewel says. Those who can do legal heavy lifting, retain existing clients and find new business are the ones valued as partners. You'll know if you aren't making the cut by your third year as an associate, Jewel says. "They may tell you that your work product or billable hours aren't what they hoped," he says. If that happens, look for work at a new firm before it's too late.

Find a Guide

Stay too long at a firm where you're not considered partnership material and your marketability will go down, Zamsky warns. "Begin networking while your marketability is still high," he suggests. "Asks friends, clients and business associates for referrals to recruiters they like and respect."

A good recruiter will help you figure out your options. Even if you're not looking to move, it's good to be in a recruiter's database and on his call list as a networking source. "The recruiter will pay you back when a great job comes up that's perfect for you," Zamsky says.

Equity or Nonequity?

If you make the partnership cut, make sure you understand what's being offered. Know the different layers of partnership at your firm, how the profit pie gets sliced and when it will be served. Many firms pay partners a draw and then make distributions to partners quarterly or annually.

Most large law firms offer two forms of partnership: equity and nonequity. An equity partnership is a true partnership, so you'll need to fund your buy-in. Equity partners own a portion of the firm's assets, including real estate, as well as its liabilities, explains Jewel.

"Firms that want to reward their associates with the partner title and prestige without diluting their ownership offer nonequity partnerships," Jewel says. With a nonequity partnership, you participate in profit sharing and gain the prestige of the partner label, but you don't own a share of the firm.

Once You're a Partner

Don't expect the big bucks to roll in the first year. Once you're a partner, if you're no longer an employee, you may have to pay your own benefits and file a partnership tax return. "The first year in partnership, your actual take-home pay can be lower than your pay the last year as a senior associate," Jewel warns.

Once you're in the club, pay extra attention to backroom deals, which committees have the most power, who supports which factions and who backstabs whom. "Because you're now a business owner, your money and your company are being affected by those factors," Zamsky adds.

Making partner means working harder than ever. "The rewards are greater, but you have to work for them," Zamsky says. "The pressure will be on to run your practice, to participate in firm management, to deliver service to clients and always to bring in new business. More is expected of you."

Counsellor's Perspective: " To become a lawyer is to take part in shaping the life of a nation and its people.

Lawyers are central figures in the life a democratic country" (Services for Legal Education, LSAC). The field of law encompasses practically every aspect of society. A law degree can lead to a rewarding legal career in which you can effect social change, set legal precedent and make a positive impact on the lives of people.

The law is an integral part of nearly every area of our lives. In the United States today there are over one million practicing attorneys. People who enter this demanding profession do so for many reasons. Lawyers work for law firms both big and small, in government agencies, legal departments of corporations, for individuals or on behalf of communities.

As an attorney you might find yourself communicating with clients and colleagues, negotiating with adversaries, drafting legal documents and preparing briefs, arguing cases before judges and juries. You could be advising corporate management on the acquisition of a competitor; representing an individual on an assault and battery case; or acting as a law school dean. If you are involved in public interest law you may represent low-income individuals while working with private firms or for non-profit organizations or be involved in organizing groups or work on law related issues in a role as advocate, educator, or mediator helping clients and communities secure or protect civil rights or social justice. Some lawyers specialize in advising corporations, some focus on taxation and estate planning; others practice public interest law. Many lawyers use their legal expertise in other fields. Some lawyers are corporate executives, bankers, legal affairs reporters, or school administrators

Most lawyers spend much of their time outside the courtroom. Their work is intellectually demanding with a great deal of time spent preparing for clients, writing briefs or legal documents, or researching of legal issues. Lawyers often specialize and may have studied other disciplines. Some lawyers do not like litigation and prefer being involved in negotiation. Others choose to make litigation, either in criminal or civil courtrooms, their focus.

The key to finding your role in the vast field of law is...first, to discover who you are.

Opportunities: The employment market for law school graduates has held steady, staying close to or above an 89% employment rate for the class of 2004. More than half (56.2%) of those who were employed obtained jobs in private practice. Employment in business was 12%, while 27.7% were in public service positions (from NALP's Jobs & JD's, Class of 2004). Lawyers held about 735,000 jobs in 2004. Private practice attorneys make up the vast majority of law jobs with approximately 3 out of 4 lawyers working in law firms or private practice. Most salaried lawyers are found in government or with corporations or non-profit organizations. About 25 percent work in government jobs, generally at the local level. In the Federal Government, lawyers can be found in many different agencies, primarily in the Departments of Justice, Treasury and Defense. Many salaried lawyers working outside of government are employed as house counsel by public utilities, banks, insurance companies, real estate agencies, manufacturing firms and other business firms and non-profit organizations.

Employment of lawyers is expected to grow about as fast as average for all occupations through 2014. Certain areas of law are expected to grow faster than others. Increase demand for legal services will be seen in areas of health care, intellectual property, venture capital, energy, elder, antitrust, international law and environmental law. In addition, greater availability and affordability of legal clinics should result in increased use of legal services by middle-income people. Some limited growth of demand for lawyers will be seen as businesses reduce costs by using accounting firms and paralegals to perform roles historically held by lawyers.

Employment of lawyers is expected to grow, primarily as a result of growth in the population and in the general level of business activities throughout the world. Employment growth of lawyers also will result from growth in demand for legal services in such areas as elder, antitrust, environmental and intellectual-property law.

The legal assistant profession is projected to grow by 33% during the coming years, according to the Bureau of Labour Statistics. As compared with other occupations, this is an above average growth rate.

Private law firms are the largest employers of law graduates, but a growing number of other organizations are beginning and continuing to hire them. These organizations include corporate legal departments, insurance companies, real estate and title insurance firms and banks.

Career Opportunities: The current industry scenario offers attractive career avenues for Indian Law Professionals. The various career options available are:

Government Services: Law Professionals can become, Law officers, Assistant Advisors, Legal Advisors, senior position in

State police/Revenue/Judicial Departments, law inspectors, legal officers in banks, judicial members of income tax, sales tax and excise departments and government advocates.

Public Prosecutor: Govt. Pleaders represent the State in courts and lends advice to officers in districts in all legal matters when called for.

Defense Service: They can work in the legal branches of defense departments. They conduct courts of inquiry and court martials of offending service personnel governed by the particular legislation acts.

Legal Advisors: They work in legal departments of firms or for private clients. Legal advisors counsel their clients as to their legal rights and obligations and suggest the appropriate course of action in business and personal matters.

Advocate: An advocate practices in the court of law. He represents a party (individual, group of individuals or a company) in the court.

Solicitor: Solicitor is a kind of advisor to clients on a variety of matters that involves law. Solicitors explain the cases of his clients to the advocates. Solicitors need never appear in court directly. Solicitor's firms undertake cases related to their area of work.

Teacher: Those with a good aptitude and taste for teaching can go for LLM and take up job in any University or Institute offering law courses to students.

Writer: An author and produces journals and legal commentaries and bring out law reports. They also write articles for magazines, newspapers etc.

Other Fields: Lawyer can also work as legal counsel and legal advisers for firms, organizations and families. They can work as trustees of various trusts, as teachers, paralegal and legal assistant, law reporters, company secretary and so on.

Eligibility: Working as a lawyer requires excellent and persuasive oral and written communication skills. Lawyers must be detail-oriented, natural negotiators who enjoy research. The basic qualification for a successful career in the industry is degree in law (LLB). However, some employers look for, other diploma courses in several disciplines of law, including Administrative Law, International Law, Labour Laws, Tax Laws and Corporate Laws in addition to the LLB degree.

Career Plans in Legal Studies

Legal careers span the full range of education requirements. Attorneys have the most specialized training, paralegal positions often require some and legal secretaries or assistants need only have a general education, preferably with some legal experience.

Legal Studies for a Career as an Attorney

A career as a lawyer requires intensive legal studies. Competition is intense for admission to most law schools, so you'll have to work hard and be committed from the outset. You'll need to complete a 4-year college degree, then three years in law school and a written bar exam before you'll qualify.

The hard work will be well rewarded in terms of job opportunities and salary. Lawyers are involved in so many areas of our lives. Almost any private or business transaction that takes place has some basis in law. Lawyers act as advisors or advocates in criminal procedures, in civil litigation and in contractual matters, including the transfer of real estate. They also act as general advisors to large businesses, banks, schools, hospitals and the government. Many lawyers work in private practice.

The central role that lawyers play in our society ensures that demand for lawyers will continue. Competition is strong, however—it's important to plan your legal studies carefully. Geographical mobility can also be an asset, especially In the early years of your legal career when gaining a wide range of experience is most important.

Legal Studies for Other Legal Careers

If the education commitment required for a career as a lawyer is a bit too daunting, consider pursuing paralegal training or entering the workforce as a legal secretary. Career prospects are similarly strong in those areas and the road there much less daunting.

The Bureau of Labour Statistics reports that legal secretaries and administrative assistants held about 4.1 million jobs in 2002, ranking them among the largest occupations in the US economy. Jobs for Paralegals are also projected to grow faster than average, as law offices try to reduce costs by assigning them tasks formerly carried out by lawyers. If you are interested in the law but not in the time and expense of law school, a paralegal or legal secretary job might be the best course.

What does it Mean to Study Legal Studies?

Legal studies careers centre on the practice of law. If you want a job protecting people's rights, lives and property, then a legal studies career will be right for you.

What Kind of Person Succeeds in Legal Studies?

If you're hooked on CourtTV, you know that most folks in legal studies careers love to argue. You'll need a great mind for detail and a mastery of the law, whether you're a legal secretary, law clerk, paralegal, or lawyer. If you're analytical, persistent, not easily intimidated and persuasive, a law-related career will fit.

Career Prospects for Legal Studies?

Paralegals are in exceptional demand right now and will be for the foreseeable future. Demand for lawyers is stable and expected to grow. Law clerks and lawyers employed by government agencies make, on the average $40,000. Lawyers who work in business, business attorneys or tax attorneys average $60,000. Lawyers in private practice average $90,000. Three out of four lawyers choose private practice.

Educating Yourself for a Legal Studies Career

A bachelor's degree is enough to work as a paralegal. Paralegals or legal assistants do many of they same jobs that lawyers do, except practice law. Lawyers need a 3-year law degree beyond the 4-year bachelor's. They intern after their second year of law school and many get their first jobs in the firms for which they interned. Of course, you'll have to pass the bar exam before you can practice law.

What's it like to Work in Legal Studies?

You'll work in law offices or libraries or in courtrooms, if you study for a legal career. You'll do research and gather evidence, whether you're a paralegal, law clerk, or lawyer. Lawyers appear in courts or before other authorities. Few in the legal field work regular hours or have 40-hour workweeks. Half work 50 hours a week or more. You'll work under lots of pressure if you're a trial lawyer.

Legal Careers—Future Outlook

The entertainment industry goes through ups and downs. Teaching jobs often rely on adequate funding. Even medicine seems volatile at times. It seems impossible to find a career with any degree of stability and certainty. However, legal jobs might come pretty close. After all, the law does not depend on foreign markets, government funding, or a fickle public. Careers in this field will always flourish regardless of the economic or political climate.

What is a Legal Career?

Legal careers apply to any jobs that focus on defending, upholding, or interpreting the law. Broadly speaking, congressional members and police officers fall into this category, but usually, one thinks of lawyers, judges and paralegals.

Legal Studies are Essential in this Field

With the exception of paralegals, you will probably need to complete some level of formal study in order to enter this industry. Your training will prepare you to wade through the pages of precedence, opinions and regulations. Legal careers usually involve a great deal of problem solving skills, outside-the-box thinking and strong attention to details. As you study, you'll also want to focus on writing skills since this will constitute a major portion of your work. Some attorneys produce 100-page reports on a regular basis. Are your current writing skills up this challenge?

Pick a Field and Go with It

Although lawyers have a reputation for courtroom antics (thanks to Perry Mason and others), litigation is only a portion of what they do. There are lawyers in every field imaginable. You enjoy sports? Provide legal counsel for your favourite team. You like country music? Represent the next big star. Every industry needs attorneys and paralegals. Just pick an area that interests you, tailor your studies appropriately and start making a difference. Whether for business deals, criminal cases, or general legal advice, there will always be demand for your services.

General Career Information

When it was first written, the Constitution was one of the most radical departures from established law that humans had ever created. Democracy, due process, fairness and equality were some of the major themes interwoven into this new and unconventional pact between the people and their government. In the many years since, upholding, defending and interpreting the Constitution has been a continuous and complicated process that has spawned a great many careers in law.

Careers in Law

From the judges on the Supreme Court to the corrections offices in the prison systems, there are legal jobs of all shapes and sizes. Most require at least a bachelors degree or some comparable level of study. Others require additional education though. To become an attorney, for example, you'll need to study in law school for at least three years and then pass a State bar. For law enforcement, you'll need to attend police academy. To become a judge, you'll need to practice as an attorney, clerk with an existing judge and work your way up the career ladder.

The Scope of your Duties

Your exact duty will depend on what type of legal career you want to pursue. For example, lawyers usually focus on interpreting the law with regards to business transactions, crimes, civil suits and contracts. Police officers focus on enforcing the law and keeping the peace. Social workers and corrections officers work on rehabilitating people and preparing them for re-entry into society. As you can see, the work is extremely varied.

Crime is Up—Crime is Down

Some years, studies suggest that crime is on the rise. Other years, studies suggest that crime is declining. Either way, expect to work hard. In times of relative peace and tranquillity, your job is to make sure that everything remains as it is. When crime starts to rise, your job is to capture and punish guilty individuals. The work never really ends.

Job Openings and Education Requirements—General Outlook: Projected job growth varies widely by education and training requirements. Legal Education is essential in getting a high-paying job. In fact allmost all of the highest paying occupations require a college degree.

Job openings stem from both employment growth and replacement needs. Replacement needs arise as workers leave occupations. Some transfer to other occupations while others retire, return to school, or quit to assume household responsibilities. Replacement needs are projected to account for 60 percent of the approximately 58 million job openings between 2000 and 2010. Thus, even occupations with little or no change in employment still may offer many job openings.

Jobs requiring an associate degree are projected to grow 32 percent over the 2000-10 period and jobs requiring a post-secondary vocational award grow 24.1 percent. Employment in occupations requiring an associate degree is projected to increase faster than any other occupational group categorized by education or training.

The largest number of job openings will be among occupations requiring short-term on-the-job training. Almost two-thirds of the projected job openings over the 2000-10 period will be in occupations that require on-the-job training and arise mostly from replacement needs. Occupations requiring work-related training are projected to increase 12.4 percent. These jobs will account for 37.3 million of the projected 57.9 million total job openings through 2010. However, many of these jobs typically offer low pay and benefits; this is more true of jobs requiring only short-term on-the-job training, which will account for 24.8 million openings.

Jobs requiring a bachelor's degree and which usually offer higher pay and benefits, will account for about 7.3 million job openings through 2010. Employment in occupations requiring at least a bachelor's degree is expected to grow 21.6 percent. Most of these openings will result from job growth.

How to Get Required Education for your Legal Career?

The right education is important for your career. Legal Education always depends on learning & going to school. You should consider that except for financial considerations, the public or private nature of a university should not be a factor in selecting a graduate programme. High quality programmes exist in both types of institutions.

Of more importance is the institution's commitment to the graduate programme. This commitment is found in its willingness to maintain a first-class faculty and to provide excellent facilities for advanced study, including libraries, laboratories, computers and other equipment.

Another important factor to consider in many disciplines is the presence of strong departments in other fields relevant to your interests so that you can have access to scholars and courses in disciplines related to your own. Selecting the right school that offers the required degree for your education is a large investment of time, money and effort. A consequent & careful evaluation of campus-based or online schools is important.

Find Schools that Meet your Interests

Go to the Legal School Directory and select the appropriate category, country or state. You will then find a list of Legal Schools matching your interests.

Gather Information from All Schools

Upon request every Legal School listed in the directory will provide you with information and details. In the directory just click on the link for each school or on the button "Complimentary Information" to request] about its educational programmes, finacial aid, admission requirement, etc.

Corporate Law

What with existing businesses growing in size and with a large number of new business enterprises opening shop each year, corporate lawyers are on a growth flight which is likely to last several decades.

Corporate law defines the rules of the game for business corporations. And corporate lawyers come into play whenever any of the parties involved cries foul – which is very, very often, indeed – to advise their corporate clients, to argue their cases in court, draw up business contracts for them and so on.

The Work: Lawyers in private practice usually work for long and irregular hours meeting clients, researching and reading about new developments, drafting contracts and agreements or preparing briefs (particularly when their cases are being tried in court). On the other hand, their salaried counterparts in the Judicial Service or those in corporate sector tend to have saner and more structured work schedules.

As a lawyer, you will end up spending a considerable part of your time at your desk. Hence, cultivating the art of 'intelligent' listening, sifting the grain from the chaff, counselling, negotiating and writing/drafting are absolutely vital for a successful lawyer. The rest will come with experience.

This is a profession where family connections still matter. Reputation is what clients go by in the legal profession and that has to be earned. It rarely comes easy or fast.

Employment Opportunities: So far the usual image we have of lawyers in India is that of an individual advocate practising his profession from small home offices and crowded cubicles and rooms – grandly called "chambers"–attached to various courts. On the advocate's desk are piles of files tied up with strings; in the courts themselves there is much pushing, shoving and crowding. The one-man legal practitioner combines court appearances in the day with client meetings and legal drafting in the evenings.

Well, the big emerging trend in law is the move away from solo legal practice to well-organised law firms comprising several or even several dozens legal eagles working together as one entity, offering a full range of legal expertise. While the number of law firms is still relatively small–and the solo legal practitioner is not about to disappear anytime soon – the trend towards law firms is unmistakable. And for those seeking careers in corporate law, it might be a good idea to look at law firms because corporate clients are more likely to deal with legal firms rather than individual lawyers. Corporations simply feel more comfortable dealing with other companies.

In case you choose to set up your own practice, you'll first need to learn the ropes under a senior advocate or, of course, in a law firm.

The legal profession has undergone a sea change over the last decade. Especially in the field of corporate law as the rules governing companies have been virtually rewritten with the opening up of the market in the last decade.

The changing face of corporate law is increasingly catering to the specific requirements of each organisation–large or small. If you can figure out the core functions of an organisation, you can tailor the generic law to suit their requirements.

Broadly speaking, corporate lawyers advise corporations on their legal rights and obligations. Very few matters actually come to the court. Multi-crore suits have changed the very complexion of the game, making corporate law a popular and lucrative career option. And with foreign law firms moving into India, salaries have begun to shoot upwards.

Demand for lawyers with experience in securities and transactions, is growing in proportion to changes in the global marketplace. While till some time ago, only major companies transacted international business, even small players are now involved in foreign transactions necessitating the services of corporate lawyers who are well-versed with International law.

In fact, with joint ventures and foreign collaborations coming in, the need for drawing up contracts and interpreting statutes and statutory provisions would increase considerably. Contract law is gaining importance, thanks to increased private sector participation and liberalisation. Securities and banking laws are also gaining importance with the advent of e-commerce.

Besides large corporate houses, some of the big accounting firms like PwC and KPMG have opened their own legal departments. And with global law firms poised to enter India, the market for corporate lawyers is on the verge of an exciting spin.

To boost your prospects further, you could even plumb for an additional qualification, *e.g.*, Company Secretaryship or a degree in HR or Industrial Relations with a working knowledge of labour laws depending on the field you wish to specialise in. Conversely, you could opt for a Master's in Business Laws from NLSIU (through distance learning mode) after your MBA or CS to boost your prospects.

If you have a flair for writing, you could initially attach yourself to the legal correspondent of a newspaper and move on to independent reporting in due course. If you are academically inclined, you could also join a reputed legal publishing house or go into teaching, research or perhaps even politics.

***Where to Study*?:** You can either study law straight after Class 12 or opt for a 3-year LLB course from a reputed university after your graduation.

Integrated 5-year Law courses (BA LLB) are offered at over 50 universities in India.

Job Responsibilities

Wondering what corporate attorneys actually do? Take a look at the job responsibilities in corporate law.

Due Diligence

Due diligence, also called document review, is the process of reviewing existing legal and business contracts of a business (including corporate documents, agreements and financial statements) for potential problems and issues prior to a proposed transaction, such as a merger or acquisition. The usual goal is to make sure that there is nothing in any of the contracts that would prohibit the sale of the company (or require a third party's consent) and to make sure that the contract will not terminate as a result of the sale. Usually, an associate will catalog the documents he or she has reviewed and write summaries of the key agreements. For example, when a merger is being considered and the acquirer hasn't had the time to read all of the contracts of the company it is considering purchasing, the lawyers will be expected to summarize the agreements. The purpose of due diligence is to give your client the clearest possible picture of the company you're examining, so the client can assess the risks and benefits of going through with the contemplated transaction.

In an initial public offering, due diligence involves reading agreements that are summarized in the deal prospectus—the document given to prospective investors summarizing important information about the company and the deal—to make sure the prospectus is correct. If you are representing an underwriter in a securities offering, you will review documents so that the underwriter can claim to have made a reasonable investigation of the issuer's statements in the offering document. Basically, it's done so that the underwriter will be able to claim the due diligence defense if there is a material misstatement in the offering document. ("I investigated and everything looked fine so don't come after me with a lawsuit because the investors lost their money.")

A junior associate does more due diligence than contract drafting because one of the best ways to learn to draft contracts is to read a lot of them. In the past, lawyers frequently travelled to the company they were reviewing and spent a few days, or however long it took, to review all the materials pertinent to the transaction. Some attorneys have noticed a trend toward cutting back on lawyer travel. Apparently, clients have discovered it's cheaper to have their own employees copy all the documents and send them to the law firm for review.

If you do travel to the company, document review is a great way to bill a lot of hours in a short period of time. You don't waste time looking for work or trying to decide whether to go to lunch with your colleagues. And no one can give you any non-billable work to do, like updating the firm's research files. The company sets aside a room for you and puts all the documents in there and you read and read and read all day. They send in breakfast and lunch and you keep reading (and billing). If the company will not let you stay there when they close for the evening, you and your colleagues will go out to dinner and stay overnight in a nice hotel, for which the client pays. If the company does let you stay in their offices, you might be there until 1 or 2 in the morning. Most lawyers find the work tedious, but it's excellent training on how to draft contracts and how to review them for pertinent information. In addition, it's a chance to get to know your colleagues really well. On rare occasions, you'll get to do a little detective work if the company has not supplied the proper documents and you have to interview employees to determine where they are.

Research and Writing Memos

Researching legal issues and writing memoranda about them is something that you should learn in law school, as research and memoranda writing are essential for all aspects of a law practice. However, this process is rarely the same in a law firm. Not only are memos in corporate law slightly less formal and slightly less constrained by citation rules, your superiors might have very different ideas about what constitutes a useful, readable document. In a corporate law department, you are looking up applicable laws and figuring out what practical impact they have on your client's decisions and actions, rather than assessing the state of the law on a certain issue in the abstract. It's important to research and write the memo as quickly and efficiently as possible without sacrificing thoroughness and accuracy. Don't be surprised if your first memorandum comes back to you bleeding with editorial suggestions. And don't let it bother you. Just make sure to incorporate all of the suggestions into your revision and ask questions if you don't understand a comment.

Contract Drafting and Review

Drafting a contract is preparing the contract from the beginning—usually starting with a form and then tailoring it to fit the needs of the parties. You put down in words what the parties have agreed to in principal as best as you can so that there are no ambiguities in the future when the same parties or others read those documents. Then you or your senior associate will go back and forth with the lawyers for the other parties to revise and refine the document until all parties are comfortable signing. Negotiation of the contract involves some compromise. Once you understand what is important to your client and what they can live with or without, you try to strike the best deal for them without risking endless delay or total breakdown of the process. It's best not to spend too much time trying to hash out issues that aren't really essential to your client.

Review of a contract is simply reading the contract to determine what the parties' rights are under its terms and whether any of the terms may be detrimental to your client. When you're starting out, the person who gives you the assignment will tell you what to look for. Ultimately, you'll have read so many agreements that you'll know what's standard and what could be disadvantageous to your client.

Formation and "Housekeeping" of Corporate Entities

As a new corporate associate, you will learn to draft basic certificates, like a certificate of incorporation. These certificates are modelled after a form, so such tasks are pretty easy. In many firms, this kind of work is also handled by experienced paralegals and those paralegals can often be a great source of knowledge for a new associate.

But before you file the papers, someone has to determine what would be the best structure of the organization, depending on the clients' goals. Examples of different legal entities include general partnerships, limited partnerships, limited liability companies, C corporations, S corporations and business trusts. Each structure offers its own legal and tax advantages and disadvantages. For example, a limited liability company, which is a hybrid between a partnership and a corporation, might provide the best tax treatment while retaining the limited liability advantage of a corporation. If the company wants to trade shares on an open market ("go public"), they have to use a standard corporate form, or C corporation, since that is what stock exchanges usually require. For a smaller company, with no immediate plans of going public, an S corporation, which, unlike the C corporation, does not pay income taxes on earnings (instead, the shareholder pays income tax on dividends), might present a better solution.

Forming the company is easy and a paralegal can usually accomplish this task, although it is a good idea for a new lawyer to learn the process and be able to review the paralegal's work. The hard part is determining what the rights of the owners are in relation to each other, including consideration of how the parties are going to get out of the relationship if they decide their goals are not being met and making sure that all of this is reflected in the company's documents, whether via a shareholders' agreement or an LLC agreement, in the articles of incorporation or bylaws, or even on the share certificates themselves. As a new lawyer, you won't be making these decisions. Either you'll be executing the instructions of another attorney, or taking a crack at the project yourself and then learning the correct reasoning when the someone else reviews your work.

After you have created a legal entity under the applicable state law, you have to continue to ensure that proper documentation is made of the company's decisions and actions so that there is a clear corporate history of regulatory compliance. Corporate "housekeeping" involves drafting and filing the necessary paperwork to maintain an entity's corporate status with the appropriate governmental entity, usually a state secretary of state's office and the Internal Revenue Service. For example, you may draft and/or review board of directors' resolutions, file tax records, maintain and prepare stock certificates and draft partnership or shareholder agreements.

Often, small companies are formed without the advice of a lawyer. When this happens, the corporate formalities, other than the filing of simple articles or certificate of incorporation might be largely ignored. This doesn't usually become a problem for the company until it becomes an acquisition target (meaning someone wants to buy the company) or when the company has expanded and wants either to start raising capital through the sale of shares or to go public. At this point, a lawyer will often have to work with the client to recreate the records that the corporation needs and may even have to amend the articles of incorporation to permit for the issuance of additional and different types of shares. You will have either to know or to research the corporate statutes of the state of the company's organization to make sure that the company is in compliance with the state's requirements.

Preparing Filings for State and Federal Agencies

It's likely that a new associate will prepare filings other than corporate organizational documents. At a minimum, an associate will usually get some exposure to corporate securities filings.

When a company wants to start selling shares to people who were not involved in the inception and management of the company, that company must share a great deal of information about itself. The kind of information the company must provide and the format in which the information must be provided depend on the level of financial sophistication of its investors. The Securities and Exchange Commission (SEC) regulates this process.

If your client wants to sell shares to all kinds of investors, they "go public" and are "traded on an open market." If they choose to do this, they will be subject to very strict disclosure requirements. The documents they have to produce are lengthy (usually over a hundred pages) and extremely number-intensive. They must comply with very specific rules and regulations. Most associates will spend some time proofreading these documents and researching applicable laws and regulations.

Working In-house

Perhaps you've heard lawyers and law students liken their search for an in-house position to a quest for the Holy Grail. What's so different about this kind of position? What do these lawyers do?

Perhaps you've heard lawyers and law students liken their search for an in-house position to a quest for the Holy Grail. What's so different about this kind of position? What do these lawyers do?

"Running a business is all about establishing relationships to help your business run," explains Daniel Bookvar. "In-house lawyers document and solidify the relationships for your organization. These lawyers create contracts with suppliers and contracts with customers and contracts with the people who keep the lights on. The most technical transactions are always outsourced. If you think of your relationship with your doctors, the in-house counsel is like the general practitioner. He's the one with the close relationship with the patient and he knows how the arm relates to the head and how the head relates to the neck bone, but he's not the guy you pull in for hand surgery."

The duties of an in-house attorney can vary greatly depending on the company and the position. Some lawyers with corporate law training end up managing litigation in addition to drafting and negotiating contracts and leases. "I get in between 9:00 and 9:30 and leave between 7:00 and 7:30," reports the in-house counsel of a New York media company. "As soon as I get in there's a phone call from a business person asking me to look at a document or asking me to help out with something. I'm constantly looking at documents, marking them up and having discussions with the business people about how long we want the term of sale to be, or how to establish pricing, or "Gee, does this really make sense given what you've told me?" I'm constantly on the phone, either with my own clients (the company's business or sales people) or with the other side. It's an incredibly interesting day."

For many attorneys, their jobs in the legal departments of corporations end up involving more than pure legal work. They also use an intuitive or acquired understanding of the business side of the company. It's often at in-house positions that an MBA or equivalent business training will prove most helpful.

The advantages of going in-house are many. Some lawyers set their own hours and can choose to work at home on occasion. Some must follow stricter schedules and be in the office between certain hours of the day. Many lawyers, particularly in smaller legal departments, feel a greater sense of autonomy than they did as associates at a law firm. Since the company has handpicked them from large law firms, the company assumes these lawyers are already well-trained and do not need to have their work constantly reviewed. This independence can be a tremendous change for an attorney coming out of a large firm. Some have described it as feeling as though they're finally back in control of their lives.

Many lawyers also thrive on the constant communication with the business people of the country. Every lawyer cites the freedom from billable hours. "The billable hour requirement," muses one in-house attorney, "just makes things worse. I think I've worked more hours straight through while in-house than while I was in a law firm. I think the pressure to constantly be billing hours in some ways slowed me down. Now I don't even think about it; I just get a lot of work done because I have a lot of different things that have to be done."

Another difference between practicing in a firm and working in-house is in the type of agreements in-house lawyers draft and negotiate. Companies aren't usually buying and selling each other every day, so in-house lawyers are no longer doing as much of the merger and acquisition work common to the training of a large law firm associate. Instead, they focus on what the company does on a daily basis, which could be selling a product or a service or pieces of real estate. In-house lawyers report that this work involves very different agreements and takes place at a completely different pace. Commercial law (the law of buying and selling goods and services) uses a more standard form contract and many of the discussions are around purely commercial terms—such as the pricing, the quantity and so on.

The Many Faces of a Bankruptcy Lawyer

A little bit corporate, a little bit litigation, a little bit employment law. Bankruptcy law allows practitioners to try on many different roles.

Bankruptcy is a haven for Geminis, sufferers of Multiple Personality Disorder and those who are a little bit country and a little bit rock-and-roll. As will be a recurring theme in this guide, no other practice allows the practitioner to try on so many different roles—and the opportunity to excel at them. Gregory Willard, head of Bryan Cave LLP's Bankruptcy, Restructuring and Creditors' Rights Client Service Group speaks for most bankruptcy practitioners when he describes bankruptcy as the "last bastion of generalists."

Bankruptcy is more than simply a series of hearings before a bankruptcy judge. It's a court-managed transactional process, with every transactional decision having a litigation counterpart and vice versa. Appropriately, bankruptcy attorneys get to play both sides of the fence, all along counseling their clients on the impact of the Code on their business decisions. And within the transactional and litigation worlds, every area of law is touched upon. In bankruptcy, "every legal relationship has to be resolved," observes New Jersey practitioner Ben Becker of Becker Meisel, LLC and those relationships run across the substantive spectrum of law. Among the most common roles that a bankruptcy attorney plays are:

Litigator: It's no surprise that in many law firms, the bankruptcy group is a sub-section of the litigation practice. Bankruptcy is a courtroom process, with a Bankruptcy Judge overseeing each bankruptcy proceeding and approving all major debtor—and some creditor—business decisions. A bankruptcy attorney spends much of her time in court, arguing for or against debtor and creditor motions and often participating in full-scale trials.

Commercial Financing: In the book cases, where the debtor continues operating as a "going concern," a primary focus of the weeks before and immediately after the bankruptcy filing is negotiating "debtor-in-possession"—or DIP—financing. These loans provide debtors with their financial lifelines, provide bankrupt corporations with sufficient funds to keep their doors open (or at least close them very slowly and carefully). Bankruptcy attorneys negotiate and draft these DIP financing agreements, fluent in the provisions of the Code and case law governing these facilities.

Mergers and Acquisitions Attorney: In this book cases revolve around asset sales, often of the entire business of the debtor. At some firms, bankruptcy attorneys work hand-in-hand with M&A attorneys in creating auctions processes (yep, you heard that— auctions, often complete with gavels) and negotiating and closing these sales; More often, the bankruptcy attorney performs all of the M&A work, including negotiating and drafting relevant purchase and ancillary agreements and overseeing the sale closings. And in all situations, it is the bankruptcy attorneys who request (and hopefully secure) the Court's approval of these sales.

Labour and Employment: Bankruptcies involve businesses and businesses have employees. Bankruptcy attorneys provide counsel in communicating with a debtor's labour force, complying with federal and state laws and stand on the front lines of negotiations of and disputes about, employee severance and retention programmes.

Uniform Commercial Code/Secured Transactions: Creditors' secured status—and whether debtors can "avoid", or void, a creditor's security interest—is an important issue in all bankruptcies and most bankruptcy attorneys acquire some knowledge of and often expertise in, the Uniform Commercial Code and other aspects of security interests.

Chief Operating Officer: In the book proceedings and the occasional the book case, the court appoints a bankruptcy trustee, often a bankruptcy attorney, to operate and/or wind-down the debtor's business. Here, the trustee gets to act as businessman and client, typically hiring another attorney to do all legal work.

And that list just scratches the surface. Bankruptcies also involve issues of tax, real estate, landlord-tenant relations, environmental law, telecommunications, securities regulation, securities fraud, anti-trust, white-collar crime, domestic relations well, the list doesn't stop, given that every bankruptcy is different and involves every aspect of the debtor's financial life. A consumer bankruptcy involving a man in the midst of a divorce will involve all sorts of family law issues; the bankruptcy of a media conglomerate will require knowledge of federal communications regulations. This sheer variety of roles makes bankruptcy a unique and exciting practice, guaranteeing nary a dull moment.

Law and Related Options

The image of black jacketed lawyers has undergone a sea change with white collar legal executives assuming a covetable profile. Law has always been one of the most important social institutions. It establishes the rules that define a person's rights and obligations and defines their enforcement. The more developed a legal order, the more demands and responsibilities are placed upon its legal professionals. Traditionally, lawyers practice before judges in Subordinate Courts, Administrative Tribunals, High Courts and the Supreme Court. Their practice pertains to mainly private matters and public matters.

Private matters are those related to the rights and obligations people have in their involvement with one another, or with Civil Law.

Public matters are those which deal with the rights and obligations people have as members of a society. Law related to public matters concerns itself specifically to criminal cases as well as to administrative, constitutional and international cases. The areas in which lawyers specialise are:

- *Criminal Law*: Criminal lawyers specialise in matters having legal implications such as theft, murder and arson which are considered offences against society or the State.
- *Real Estate:* Real estate lawyers deal with cases involving property.
- *Civil Law:* Civil Law operates in matters of property, contract and commercial issues, inheritance, family and corporations.
- *Tax Law:* Tax lawyers handle cases concerning income tax, wealth tax, gift tax, estate duty, etc.
- *International Law:* A set of international legal rules are observed by all the nations in relation to one another.
- *Labour Law:* Labour lawyers specialise in solving the problems arising between the management and the labour force in any organisation.
- *Others*: Other branches include family law, constitutional law, corporate law, excise law, customs and transportation. Globalisation of businesses has created a need for legal

compliance, which has created a special niche for legal eagles dealing in all matters concerning business.

Law Enforcement Career

According to the National Centre for Women and Policing, women in law enforcement constitute about 15 % of the workforce. While there are many interesting avenues in law enforcement and criminal justice, not many women are aware of the requirements and whether they would suit the bill. Check out more on *law enforcement careers* for women and how you can go about looking for one.

Law Enforcement Career: While traditionally it was viewed that women are not suited for law enforcement jobs, it has been seen in recent decades that women in such careers can play a vital role in law enforcement. Women police officers might be particularly effective in sensitive cases such as child molestation or sexual assault. Other than policing jobs, women with professional and educational backgrounds in criminology, criminal justice, social work or other liberal arts can look for career options in criminal justice wings.

Career Opportunities: The FBI offers law enforcement careers for women in roles as diverse as linguists to FBI special agents. In this age of terrorism and foreign counterintelligence, exceptional foreign language skills can land you with an exciting career opportunity. With the necessary qualifications and skills, you can be considered for working on matters such as cyber crime, white collar crime, financial crime and fugitive and drug-trafficking issues.

Secret service jobs can involve careers with the technical, administrative or training wings. Special agents are assigned to cases of importance such as protection of the President, Vice President, former Presidents and their families as well of heads of foreign states visiting America. Security for the White House, Treasury Building also comes within this. Police jobs encompass careers with city police, state police or the federal police. Local law enforcement jobs include those of Corrections officer, Fire department and opportunities with the Sheriff department. Federal jobs can be sought in the following agencies:

- Department of Agriculture
- Department of Defense
- Department of Justice
- Department of the Treasury
- US Postal Service
- Department of Commerce
- Drug Enforcement Administration
- US. Marshals Service
- Federal Bureau of Prisons
- Internal Revenue Service
- Bureau of Alcohol, Tobacco and Firearms
- United States Customs Service
- Immigration and Naturalization Service.

Securing Law Enforcement Jobs: You can look out for law enforcement jobs by keeping on the watch for police recruiting efforts. Local newspapers are a good sourcing spot. Check out online services that post openings for law enforcement jobs. The recruitment process for law enforcement jobs involves written test and physical agility test.

Background investigation is done on probable recruits. Finally there is an interview that tests how you handle pressure and attempts to look at your attitudes. Women looking for law enforcement jobs are expected to meet the same physical, academic and psychological standards as those of their male counterparts.

Women in Law Enforcement: Evidence reveals that women police officers tend to be more effective when it comes to dealing with potentially violent situations. It has been noticed that women who have had a law enforcement official in their family are more likely to consider a career in law enforcement.

While some women tend to prefer administrative assignments, others thrive on the thrill of patrolling, criminal tracking, investigative work and policing. There are instances when women have faced discrimination, harassment and intimidation by coworkers and others at work.

Fraud Investment Lawyer

A *fraud investment lawyer* represents clients hurt by investment fraud and negligence or anyone who has suffered an investment loss. Fraud investment can cost someone his/her IRA savings, college fund for children, retirement money or even pension plan provided for employees. Under such trying circumstances, the investor suspects the financial advisor, stock broker, or insurance agent or someone who has made them invest the hard earned money. It is in this context that professional hand of an experienced legal representation becomes imperative for the investor.

Investment Fraud: While exploring such investment frauds, one can come across several reasons and factors that are responsible for turning investor dreams into nightmares:

- Questionable and suspicious practices by those involved in the sectors of investment, banking, brokerage and corporate arenas play a major role in causing investors to suffer economic pain.
- Corporations indulging in overly aggressive and fraudulent accounting practices are also a cause for such investment frauds.
- In a system where commission and financial gains thrive, some stockbrokers with conflicting allegiances recommend the investor to invest in stocks that they themselves feel are poor choices.

Once an investor has lost money due to fraud, there are two ways by which a fraud investment lawyer helps them recover their monies.

- Lawsuit or arbitration.
- Criminal prosecution.

In the US, there are several federal statutes available to fraud victims, depending on the type of fraud perpetrated. Skills required of a fraud investment lawyer

- A fraud investment lawyer should be talented, aggressive and a professional in every way.
- A fraud investment lawyer should have strong legal capabilities. He/she should be well-versed in complex

business transactions, financings and litigations, wide variety of business and securities matters. Legal matters should be handled with highest level of competence.

- The fraud investment lawyer should be a seasoned strategist and should be able to navigate around the negotiating table as well as the courtroom.
- A fraud investment lawyer should be able to provide prompt personal attention and impeccable legal services, capable of handling settlement negotiations, mediations, arbitrations and litigations. The lawyer should understand that his/her investor client is in distress and needs individualized attention and treatment. The fraud investment lawyer should appreciate that the client has spent many years earning and accumulating the money they have invested and should be dedicated to help them recover it.
- Experience is an extremely important factor in determining which fraud investment lawyer would be the most helpful in cases of claim or dispute. He/she should have sufficient practice that involves representing individuals and businesses who have suffered financial losses due to fraudulent investment schemes sold by unscrupulous or inept investment promoters, stockbrokers, investment advisors or insurance salesmen.
- The fraud investment lawyer should be a reliable and experienced securities litigation attorney. He/she should be able to provide professional stewardship and consideration for the client's finances and security of investment. Mismanagement, mishandling and misconduct on the part of these professionals tell significantly on the client's investment security. It can also destroy the clients' trust.
- A good investment fraud lawyer always works to learn the rules and regulations that govern the behaviour of investment industry professionals. It is essential for an investment fraud lawyer to understand the industry and its representatives to enable determine the merits of the claim or dispute of the investor client.

- A fraud investment lawyer should have public service motto as an integral part of himself/ his firm. A fraud investment lawyer should take the onus onto himself/ herself to educate the public and other lawyers about investment fraud, elder fraud and elder law.

Scope for a Fraud Investment Lawyer: On many occasions, the investors do not notice the arbitration clause in their account forms and this clause has important ramifications. The limitation of available avenues of recourse open to the dissatisfied investor is not understood by the investor at the time of investment. It is only many months' later that they realize that they have been misguided.

Sometimes an arbitration process is set forth as this is quicker and less expensive than court action. Under all or any of the above circumstances, the services of a talented and aggressive legal professional become important for the common man to recover the losses.

Tips for the investor to protect against investment fraud:

- Do not go by unsolicited financial advice. This is a simple method to gain the investor's confidence by which the advisor earns fees and commissions by investing the other person's money.
- Ensure that the firm/person is legitimate and confirm that the financial advisor is registered or of good standing.
- A stock broker or a brokerage firm can be liable to the investor if the investor has sustained losses in the stock market based upon investments that just were 'unsuitable recommendations'. It is imperative that the stock broker or other investment advisor is chosen after careful consideration.
- Sometimes an investment would sound 'too good to be true'. It probably is. There are several unscrupulous companies who try to entice investors with promises or returns as high as 50 or even 500 %.
- It is always advisable never to divulge personal financial information like credit card account or bank account numbers or balances over the telephone unless and until one is sure that the caller is not bogus but legitimate.

- Always insist and read the printed prospectus carefully before any investment is made.
- It is always sensible to consult ones' attorney, accountant or financial advisor to explain the terms of investment.
- Insist on receiving regular statement and monitor the activity in your account regularly.
- It is better to keep records relating to one's investment including notes of conversations one have had with brokers, salespersons, financial advisors and others.

Judges, Magistrates and Other Judicial Workers

Significant Points: A bachelor's degree and work experience are the minimum requirements for a judgeship or magistrate position, but most workers filling these positions also have law degrees.

Overall employment is projected to grow about as fast as the average, but varies by occupational specialty.

Judges and magistrates are expected encounter competition for jobs because of the prestige associated with serving on the bench.

Nature of the Work: Judges, magistrates and other judicial workers apply the law and oversee the legal process in courts according to local, State and Federal statutes. They preside over cases concerning every aspect of society, from traffic offenses to disputes over the management of professional sports to issues concerning the rights of huge corporations. All judicial workers must ensure that trials and hearings are conducted fairly and that the court safeguards the legal rights of all parties involved.

The most visible responsibility of judges is presiding over trials or hearings and listening as attorneys represent the parties present. Judges rule on the admissibility of evidence and the methods of conducting testimony and they may be called on to settle disputes between opposing attorneys. Also, they ensure that rules and procedures are followed and, if unusual circumstances arise for which standard procedures have not been established, judges interpret the law to determine the manner in which the trial will proceed.

Judges often hold pretrial hearings for cases. They listen to allegations and determine whether the evidence presented merits a trial. In criminal cases, judges may decide that persons charged with crimes should be held in jail pending trial, or they may set conditions for their release. In civil cases, they occasionally impose restrictions on the parties until a trial is held.

In many trials, juries are selected to decide guilt or innocence in criminal cases or liability and compensation in civil cases. Judges instruct juries on applicable laws, direct them to deduce the facts from the evidence presented and hear their verdict. When the law does not require a jury trial or when the parties waive their right to a jury, judges decide cases. In such instances, the judge determines guilt in criminal cases and imposes sentences; in civil cases, the judge awards relief—such as compensation for damages—to the parties to the lawsuit, called litigants. Judges also work outside the courtroom in their chambers or private offices. There, judges read documents on pleadings and motions, research legal issues, write opinions and oversee the court's operations. In some jurisdictions, judges also manage the courts' administrative and clerical staff.

Judges' duties vary according to the extent of their jurisdictions and powers. *General trial court judges* of the Federal and State court systems have jurisdiction over any case in their system. They usually try civil cases transcending the jurisdiction of lower courts and all cases involving felony offenses. Federal and State *appellate court judges*, although few in number, have the power to overrule decisions made by trial court or *administrative law judges*; appellate court judges exercise their power if they determine that legal errors were made in a case or if legal precedent does not support the judgment of the lower court. Appellate court judges rule on a small number of cases and rarely have direct contact with litigants. Instead, they usually base their decisions on lower court records and on lawyers' written and oral arguments.

Many State court judges preside in courts whose jurisdiction is limited by law to certain types of cases. A variety of titles are assigned to these judges; among the most common are *municipal court judge*, *county court judge*, *magistrate* and *justice of the peace*. Traffic violations, misdemeanors, small-claims cases and pretrial hearings constitute the bulk of the work of State court judges, but some States allow these judges to handle cases involving domestic relations, probate, contracts and other selected areas of the law.

Administrative law judges, sometimes called *hearing officers* or *adjudicators*, are employed by government agencies to make determinations for administrative agencies. These judges make decisions, for example, on a person's eligibility for various Social Security or workers' compensation benefits, on protection of the environment, on the enforcement of health and safety regulations, on employment discrimination and on compliance with economic regulatory requirements.

Arbitration, mediation and conciliation—collectively called appropriate dispute resolution (ADR)—are alternative processes that can be used to settle disputes between parties. All ADR hearings are private and confidential and the processes are less formal than a court trial. If no settlement is reached through ADR, no statements made during the proceedings are admissible as evidence in any subsequent litigation.

There are two types of arbitration—compulsory and voluntary. During compulsory arbitration, opposing parties submit their dispute to one or more impartial persons, called arbitrators, for a final and nonbinding decision. Either party may reject the ruling and request a trial in court. Voluntary arbitration is a process in which opposing parties choose one or more arbitrators to hear their dispute and submit a final, binding decision. Arbitrators usually are attorneys or business persons with expertise in a particular field. The parties identify, in advance, the issues to be resolved by arbitration, the scope of the relief to be awarded and many of the procedural aspects of the process.

Mediation, or neutral evaluation, involves an attempt by the parties to resolve their dispute with the aid of a neutral third party. This process generally is used when the parties wish to preserve their relationship. A mediator may offer suggestions, but resolution of the dispute rests with the parties themselves. Mediation proceedings also are confidential and private. If the parties are unable to reach a settlement, they are free to pursue other options. The parties usually decide in advance how they will contribute to the cost of mediation. However, many mediators volunteer their services, or they may be court staff. Courts ask that voluntary mediators provide their services at the lowest possible rate and that parties split the cost. Depending on the type of case, court-referred community mediation centres may charge a small fee to the parties involved in mediation.

Conciliation, or facilitation, is similar to mediation. The conciliator's role is to guide the parties to a settlement. The parties must decide in advance whether they will be bound by the conciliator's recommendations; they generally share equally in the cost of the conciliation.

Working Conditions: Judges, magistrates and other judicial workers do most of their work in offices, law libraries and courtrooms. Work in these occupations presents few hazards, although sitting in the same position in the courtroom for long periods can be tiring. Most judges wear robes when they are in a courtroom. Judges typically work a standard 40-hour week, but many work more than 50 hours per week. Some judges with limited jurisdiction are employed part time and divide their time between their judicial responsibilities and other careers.

Arbitrators, mediators and conciliators usually work in private offices or meeting rooms; no public record is made of the proceedings.

Training, Other Qualifications and Advancement: A bachelor's degree and work experience usually constitute the minimum requirements for a judgeship or magistrate position. A number of lawyers become judges and most judges have first been lawyers. In fact, Federal and State judges usually are required to be lawyers. About 40 States allow non lawyers to hold limited-jurisdiction judgeships, but opportunities are better for those with law experience. Federal administrative law judges must be lawyers and pass a competitive examination administered by the US Office of Personnel Management. Some

State administrative law judges and other hearing officials are not required to be lawyers.

Federal administrative law judges are appointed by various Federal agencies, with virtually lifetime tenure. Federal magistrate judges are appointed by district judges—the life-tenured Federal judges of district courts—to serve in a US district court for 8 years. A part-time Federal magistrate judge's term of office is 4 years.

Some State judges are appointed, but the remainder are elected in partisan or nonpartisan State elections. Many State and local judges serve fixed renewable terms ranging from 4 or 6 years for some trial court judgeships to as long as 14 years or even life for other trial or appellate court judgeships. Judicial nominating commissions, composed of members of the bar and the public, are used to screen candidates for judgeships in many States and for some Federal judgeships.

All States have some type of orientation for newly elected or appointed judges. The Federal Judicial Centre, American Bar Association, National Judicial College and National Centre for State Courts provide judicial education and training for judges and other judicial-branch personnel. General and continuing education courses usually last from a few days to 3 weeks in length. More than half of all States, as well as Puerto Rico, require judges to enroll in continuing education courses while serving on the bench.

Training and education requirements for arbitrators, mediators and conciliators differ from those for judges. Mediators who practice in State-funded or court-funded mediation programmes usually must meet specific training or experience standards, which vary by State and court.

In most States, individuals who offer private mediation services do not need a license, certification, or specific coursework; however, many private mediators and most of those affiliated with mediation organizations and programmes have completed mediation training and agreed to comply with certain ethical standards. For example, the American Arbitration Association (AAA) requires mediators listed on its mediation panel to complete an AAA training course, receive recommendations from the trainers and complete an apprenticeship.

Training for arbitrators, mediators and conciliators is available through independent mediation programmes, national and local mediation membership organizations and post-secondary schools. In 2004, 16 colleges or universities in the United States offered master's degrees in dispute resolution or conflict management and 2 offered doctoral degrees. Many more schools offer conflict-management specializations within other degree programmes. Degrees in public policy, law and related fields also provide good background for prospective arbitrators, mediators and conciliators.

Employment: Judges, magistrates and other judicial workers held 47,000 jobs in 2004. Judges, magistrates and magistrate judges held 27,000 jobs, all in State and local governments. Administrative law judges, adjudicators and hearing officers held about 16,000 jobs; 52 percent in State governments, 29 percent Federal Government and 20 percent in local governments. Arbitrators, mediators and conciliators held another 5,200 jobs. Approximately 40 percent worked for State and local governments. The remainder worked for labour organizations, law offices, insurance carriers and other private companies and for organizations that specialize in providing dispute resolution services.

Job Outlook: Overall employment of judges, magistrates and other judicial workers is projected to about as fast as average for all occupations through 2014. Budgetary pressures at all levels of government will hold down the hiring of judges, despite rising caseloads, particularly in Federal courts. Most job openings will arise as judges retire. However, additional openings will occur when new judgeships are authorized by law or when judges are elevated to higher judicial offices.

Public concerns about crime and safety, as well as a public willingness to go to court to settle disputes, should spur demand for judges. Both the quantity and the complexity of judges' work have increased because of developments in information technology, medical science, electronic commerce and globalization. The prestige associated with serving on the bench will ensure continued competition for judge and magistrate positions. However, a growing number of judges and candidates for judgeships are choosing to forgo the bench and work in the private sector, where pay is significantly higher. This movement may lessen the competition somewhat. Becoming a judge often is difficult because judicial candidates must compete with other qualified people and because they frequently must gain political support to be elected or appointed and getting that support can be expensive.

Employment of arbitrators, mediators and conciliators is expected to grow about as fast as the average for all occupations through 2014. Many individuals and businesses try to avoid litigation, which can involve lengthy delays, high costs, unwanted publicity and ill will. Arbitration and other alternatives to litigation usually are faster, less expensive and more conclusive, spurring demand for the services of arbitrators, mediators and conciliators. Administrative law judges also are expected to experience average growth in employment.

Earnings: Judges, magistrate judges and magistrates had median annual earnings of $93,070 in May 2004. The middle 50 percent earned between $54,140 and $124,400. The top 10 percent earned more than $141,750, while the bottom 10 percent earned less than $29,920. Median annual earnings in the industries employing the largest numbers of judges, magistrate judges and magistrates in May 2004 were $111,810 in State government and $65,800 in local government. Administrative law judges, adjudicators and hearing officers earned a median of $68,930 and arbitrators, mediators and conciliators earned a median of $54,760.

In the Federal court system, the Chief Justice of the US Supreme Court earned $208,100 in 2005 and the Associate Justices earned $199,200. Federal court of appeals judges earned $171,800 a year, while district court judges had salaries of $162,100, as did judges in the Court of Federal Claims and the Court of International Trade. Federal judges with limited jurisdiction, such as magistrates and bankruptcy court judges, had salaries of $149,132.

According to a 2004 survey by the National Centre for State Courts, salaries of chief justices of State high courts averaged $130,461 and ranged from $95,000 to $191,483. Annual salaries of associate justices of the State highest courts averaged $126,159 and ranged from $95,000 to $175,575. Salaries of State intermediate appellate court judges averaged $122,682 and ranged from $94,212 to $164,604. Salaries of State judges of general jurisdiction trial courts averaged $113,504 and ranged from $88,164 to $158,100.

Most salaried judges are provided health, life and dental insurance; pension plans; judicial immunity protection; expense accounts; vacation, holiday and sick leave; and contributions to retirement plans made on their behalf. In many States, judicial compensation committees, which make recommendations on the amount of salary increases, determine judicial salaries. States without commissions have statutes that regulate judicial salaries, link judicial salaries to the increases in pay for Federal judges, or adjust annual pay according to the change in the Consumer Price Index, calculated by the US Bureau of Labour Statistics.

Lawyers

Significant Points: Competition for job openings should be keen because of the large number of students graduating from law school each year.

Formal requirements to become a lawyer generally include a 4-year college degree, 3 years of law school and passing a written bar examination; however, some requirements may vary by State.

Competition for admission to most law schools is intense.

About 3 out of 4 lawyers practiced privately, either as partners in law firms or in solo practices.

Nature of the Work: The legal system affects nearly every aspect of our society, from buying a home to crossing the street. Lawyers form the backbone of this vital system, linking it to society in numerous ways. For that reason, they hold positions of great responsibility and are obligated to adhere to a strict code of ethics.

Lawyers, also called *Attorneys*, act as both advocates and advisors in our society. As advocates, they represent one of the parties in criminal and civil trials by presenting evidence and arguing in court to support their client. As advisors, lawyers counsel their clients concerning their legal rights and obligations and suggest particular courses of action in business and personal matters. Whether acting as an advocate or an advisor, all attorneys research the intent of laws and judicial decisions and apply the law to the specific circumstances faced by their client.

The more detailed aspects of a lawyer's job depend upon his or her field of specialization and position. Although all lawyers are licensed to represent parties in court, some appear in court more frequently than others. Trial lawyers, who specialize in trial work, must be able to think quickly and speak with ease and authority. In addition, familiarity with courtroom rules and strategy is particularly important in trial work. Still, trial lawyers spend the majority of their time outside the courtroom, conducting research, interviewing clients and witnesses and handling other details in preparation for a trial.

Lawyers may specialize in a number of areas, such as bankruptcy, probate, international, or elder law. Those specializing in environmental law, for example, may represent interest groups, waste disposal companies, or construction firms in their dealings with the US Environmental Protection Agency (EPA) and other Federal and State agencies.

These lawyers help clients prepare and file for licenses and applications for approval before certain activities may occur. In addition, they represent clients' interests in administrative adjudications.

Some lawyers specialize in the growing field of intellectual property, helping to protect clients' claims to copyrights, artwork under contract, product designs and computer programmes. Still other lawyers advise insurance companies about the legality of insurance transactions, guiding the company in writing insurance policies to conform with the law and to protect the companies from unwarranted claims. When claims are filed against insurance companies, these attorneys review the claims and represent the companies in court.

Most lawyers are in private practice, concentrating on criminal or civil law. In criminal law, lawyers represent individuals who have been charged with crimes and argue their cases in courts of law. Attorneys dealing with civil law assist clients with litigation, wills, trusts, contracts, mortgages, titles and leases. Other lawyers handle only public-interest cases—civil or criminal—which may have an impact extending well beyond the individual client.

Lawyers are sometimes employed full time by a single client. If the client is a corporation, the lawyer is known as "house counsel" and usually advises the company concerning legal issues related to its business activities. These issues might involve patents, government regulations, contracts with other companies, property interests, or collective bargaining agreements with unions.

A significant number of attorneys are employed at the various levels of government. Lawyers who work for State attorneys general, prosecutors, public defenders and courts play a key role in the criminal justice system. At the Federal level, attorneys investigate cases for the US Department of Justice and other agencies. Government lawyers also help develop programmes, draft and interpret laws and legislation, establish enforcement procedures and argue civil and criminal cases on behalf of the government.

Other lawyers work for legal aid societies—private, non-profit organizations established to serve disadvantaged people. These lawyers generally handle civil, rather than criminal, cases. A relatively small number of trained attorneys work in law schools. Most are faculty members who specialize in one or more subjects; however, some serve as administrators. Others work full time in non-academic settings and teach part time.

Lawyers are increasingly using various forms of technology to perform their varied tasks more efficiently. Although all lawyers continue to use law libraries to prepare cases, some supplement conventional printed sources with computer sources, such as the Internet and legal databases. Software is used to search this legal literature automatically and to identify legal texts relevant to a specific case. In litigation involving many supporting documents, lawyers may use computers to organize and index material. Lawyers also utilize electronic filing, videoconferencing and voice-recognition technology to share information more effectively with other parties involved in a case.

Working Conditions: Lawyers do most of their work in offices, law libraries and courtrooms. They sometimes meet in clients' homes or places of business and, when necessary, in hospitals or prisons. They may travel to attend meetings, gather evidence and appear before courts, legislative bodies and other authorities.

Salaried lawyers usually have structured work schedules. Lawyers who are in private practice may work irregular hours while conducting research, conferring with clients, or preparing briefs during non-office hours. Lawyers often work long hours and of those who regularly work full time, about half work 50 hours or more per week. They may face particularly heavy pressure when a case is being tried. Preparation for court includes keeping abreast of the latest laws and judicial decisions.

Although legal work generally is not seasonal, the work of tax lawyers and other specialists may be an exception. Because lawyers in private practice often can determine their own workload and the point at which they will retire, many stay in practice well beyond the usual retirement age.

Training, Other Qualifications and Advancement: To practice law in the courts of any State or other jurisdiction, a person must be licensed, or admitted to its bar, under rules established by the jurisdiction's highest court. All States require that applicants for admission to the bar pass a written bar examination; most States also require applicants to pass a separate written ethics examination. Lawyers who have been admitted to the bar in one State occasionally may be admitted to the bar in another without taking an examination if they meet the latter jurisdiction's standards of good moral character and a specified period of legal experience. In most cases, however, lawyers must pass the bar examination in each State in which they plan to practice. Federal courts and agencies set their own qualifications for those practicing before or in them.

To qualify for the bar examination in most States, an applicant usually must earn a college degree and graduate from a law school accredited by the American Bar Association (ABA) or the proper State authorities. ABA accreditation signifies that the law school—particularly its library and faculty—meets certain standards developed to promote quality legal education. As of 2005, there were 191 ABA-accredited law schools; others were approved by State authorities only. With certain exceptions, graduates of schools not approved by the ABA are restricted to taking the bar examination and practicing in the State or other jurisdiction in which the school is located; most of these schools are in California. In 2005, seven States—California, Maine, New York, Vermont, Virginia, Washington and Wyoming—accepted the study of law in a law office as qualification for taking the bar examination; three jurisdictions—California, the District of Columbia and New Mexico—now accept the study of law by correspondence. Several States require registration and approval of students by the State Board of Law Examiners, either before the students enter law school or during their early years of legal study.

Although there is no nationwide bar examination, 48 States, the District of Columbia, Guam, the Northern Mariana Islands, Puerto Rico and the Virgin Islands require the 6-hour Multistate Bar Examination (MBE) as part of the overall bar examination; the MBE is not required in Louisiana or Washington. The MBE covers a broad range of issues and sometimes a locally prepared State bar examination is given in addition to it. The 3-hour Multistate Essay Examination (MEE) is used as part of the bar examination in several States. States vary in their use of MBE and MEE scores.

Many States also require Multistate Performance Testing (MPT) to test the practical skills of beginning lawyers. Requirements vary by State, although the test usually is taken at the same time as the bar exam and is a one-time requirement.

The required college and law school education usually takes 7 years of full-time study after high school—4 years of undergraduate study, followed by 3 years of law school. Law school applicants must have a bachelor's degree to qualify for admission. To meet the needs of students who can attend only part time, a number of law schools have night or part-time divisions, which usually require 4 years of study; about 1 in 10 graduates from ABA-approved schools attended part time.

Although there is no recommended "prelaw" major, prospective lawyers should develop proficiency in writing and speaking, reading, researching, analyzing and thinking logically—skills needed to succeed both in law school and in the profession. Regardless of major, a multidisciplinary background is recommended. Courses in English, foreign languages, public speaking, government, philosophy, history, economics, mathematics and computer science, among others, are useful. Students interested in a particular aspect of law may find related courses helpful. For example, prospective patent lawyers need a strong background in engineering or science and future tax lawyers must have extensive knowledge of accounting.

Acceptance by most law schools depends on the applicant's ability to demonstrate an aptitude for the study of law, usually through good undergraduate grades, the Law School Admission Test (LSAT), the quality of the applicant's undergraduate school, any prior work experience and sometimes, a personal interview. However, law schools vary in the weight they place on each of these and other factors.

All law schools approved by the ABA require applicants to take the LSAT. Nearly all law schools require applicants to have certified transcripts sent to the Law School Data Assembly Service, which then submits the applicants' LSAT scores and their standardized records of college grades to the law schools of their choice. Both this service and the LSAT are administered by the Law School Admission Council. Competition for admission to many law schools—especially the most prestigious ones—generally is intense, with the number of applicants greatly exceeding the number that can be admitted.

During the first year or year and a half of law school, students usually study core courses, such as constitutional law, contracts, property law, torts, civil procedure and legal writing. In the remaining time, they may elect specialized courses in fields such as tax, labour, or corporate law. Law students often acquire practical experience by participating in school-sponsored legal clinic activities; in the school's moot court competitions, in which students conduct appellate arguments; in practice trials under the supervision of experienced lawyers and judges; and through research and writing on legal issues for the school's law journal.

A number of law schools have clinical programmes in which students gain legal experience through practice trials and projects under the supervision of practicing lawyers and law school faculty. Law school clinical programmes might include work in legal aid clinics, for example, or on the staff of legislative committees. Part-time or summer clerkships in law firms, government agencies and corporate legal departments also provide valuable experience. Such training can lead directly to a job after graduation and can help students decide what kind of practice best suits them. Clerkships also may be an important source of financial aid.

In 2004, law school graduates in 52 jurisdictions were required to pass the Multistate Professional Responsibility Examination (MPRE), which tests their knowledge of the ABA codes on professional responsibility and judicial conduct. In some States, the MPRE may be taken during law school, usually after completing a course on legal ethics.

Law school graduates receive the degree of *Juris Doctor* (J. D.) as the first professional degree. Advanced law degrees may be desirable for those planning to specialize, research, or teach. Some law students pursue joint degree programmes, which usually require an additional semester or year of study. Joint degree programmes are offered in a number of areas, including law and business administration or public administration.

After graduation, lawyers must keep informed about legal and non-legal developments that affect their practices. Currently, 40 States and jurisdictions mandate continuing legal education (CLE). Many law schools and State and local bar associations provide continuing education courses that help lawyers stay abreast of recent developments. Some States allow CLE credits to be obtained through participation in seminars on the Internet. The practice of law involves a great deal of responsibility. Individuals planning careers in law should like to work with people and be able to win the respect and confidence of their clients, associates and the public. Perseverance, creativity and reasoning ability also are essential to lawyers, who often analyze complex cases and handle new and unique legal problems.

Most beginning lawyers start in salaried positions. Newly hired salaried attorneys usually start as associates and work with more experienced lawyers or judges. After several years of gaining more responsibilities, some lawyers are admitted to partnership in their firm or go into practice for themselves. Some experienced lawyers are nominated or elected to judgeships. Others become full-time law school faculty or administrators; a growing number of these lawyers have advanced degrees in other fields as well.

Some attorneys use their legal training in administrative or managerial positions in various departments of large corporations. A transfer from a corporation's legal department to another department often is viewed as a way to gain administrative experience and rise in the ranks of management.

Employment: Lawyers held about 735,000 jobs in 2004. Approximately 3 out of 4 lawyers practiced privately, either as partners in law firms or in solo practices. Most salaried lawyers held positions in government or with corporations or non-profit organizations. The greatest number of lawyers working in government were employed at the local level. In the Federal Government, lawyers work for many different agencies, but are concentrated in the Departments of Justice, Treasury and Defense. Many salaried lawyers working outside of government are employed as house counsel by public utilities, banks, insurance companies, real estate agencies, manufacturing firms and other business firms and non-profit organizations. Some also have part-time independent practices, while others work part time as lawyers and full time in another occupation.

Job Outlook: Employment of lawyers is expected to grow about as fast as average for all occupations through 2014, primarily as a result of growth in the population and in the general level of business activities. Job growth among lawyers also will result from increasing demand for legal services in such areas as health care, intellectual property, venture capital, energy, elder, antitrust and environmental law. In addition, the wider availability and affordability of legal clinics should result in increased use of legal services by middle-income people. However, growth in demand for lawyers will be limited as businesses, in an effort to reduce costs, increasingly use large accounting firms and paralegals to perform some of the same functions that lawyers do. For example, accounting firms may provide employee-benefit counseling, process documents, or handle various other services previously performed by a law firm. Also, mediation and dispute resolution increasingly are being used as alternatives to litigation.

Competition for job openings should continue to be keen because of the large number of students graduating from law school each year. Graduates with superior academic records from highly regarded law schools will have the best job opportunities. Perhaps as a result of competition for attorney positions, lawyers are increasingly finding work in non-traditional areas for which legal training is an asset, but not normally a requirement—for example, administrative, managerial and business positions in banks, insurance firms, real estate companies, government agencies and other organizations. Employment opportunities are expected to continue to arise in these organizations at a growing rate.

As in the past, some graduates may have to accept positions in areas outside of their field of interest or for which they feel over-qualified. Some recent law school graduates who have been unable to find permanent positions are turning to the growing number of temporary staffing firms that place attorneys in short-term jobs until they are able to secure full-time positions. This service allows companies to hire lawyers on an "as-needed" basis and permits beginning lawyers to develop practical skills while looking for permanent positions.

Because of the keen competition for jobs, a law graduate's geographic mobility and work experience assume greater importance. The willingness to relocate may be an advantage in getting a job, but to be licensed in another State, a lawyer may have to take an additional State bar examination. In addition, employers are increasingly seeking graduates who have advanced law degrees and experience in a specialty, such as tax, patent, or admiralty law.

Employment growth for lawyers will continue to be concentrated in salaried jobs, as businesses and all levels of government employ a growing number of staff attorneys and as employment in the legal services industry grows. Most salaried positions are in urban areas where government agencies, law firms and big corporations are concentrated. The number of self-employed lawyers is expected to decrease slowly, reflecting the difficulty of establishing a profitable new practice in the face of competition from larger, established law firms. Moreover, the growing complexity of law, which encourages specialization, along with the cost of maintaining up-to-date legal research materials, favours larger firms.

For lawyers who wish to work independently, establishing a new practice will probably be easiest in small towns and expanding suburban areas. In such communities, competition from larger, established law firms is likely to be less keen than in big cities and new lawyers may find it easier to become known to potential clients.

Salaries of experienced attorneys vary widely according to the type, size and location of their employer. Lawyers who own their own practices usually earn less than those who are partners in law firms. Lawyers starting their own practice may need to work part time in other occupations to supplement their income until their practice is well established.

Most salaried lawyers are provided health and life insurance and contributions are made to retirement plans on their behalf. Lawyers who practice independently are covered only if they arrange and pay for such benefits themselves.

Paralegals and Legal Assistants

Significant Points: About 7 out of 10 work for law firms; others work for corporate legal departments and government agencies.

Most entrants have an associate's degree in paralegal studies, or a bachelor's degree coupled with a certificate in paralegal studies.

Employment is projected to grow much faster than average, as employers try to reduce costs by hiring paralegals to perform tasks formerly carried out by lawyers.

Competition for jobs should continue; experienced, formally trained paralegals should have the best employment opportunities.

Nature of the Work: While lawyers assume ultimate responsibility for legal work, they often delegate many of their tasks to paralegals. In fact, paralegals—also called legal assistants—are continuing to assume a growing range of tasks in the Nation's legal offices and perform many of the same tasks as lawyers. Nevertheless, they are still explicitly prohibited from carrying out duties that are considered to be the practice of law, such as setting legal fees, giving legal advice and presenting cases in court.

One of a paralegal's most important tasks is helping lawyers prepare for closings, hearings, trials and corporate meetings. Paralegals investigate the facts of cases and ensure that all relevant information is considered. They also identify appropriate laws, judicial decisions, legal articles and other materials that are relevant to assigned cases. After they analyze and organize the information, paralegals may prepare written reports that attorneys use in determining how cases should be handled. Should attorneys decide to file lawsuits on behalf of clients, paralegals may help prepare the legal arguments, draft pleadings and motions to be filed with the court, obtain affidavits and assist attorneys during trials. Paralegals also organize and track files of all important case documents and make them available and easily accessible to attorneys.

In addition to this preparatory work, paralegals perform a number of other vital functions. For example, they help draft contracts, mortgages, separation agreements and instruments of trust. They also may assist in preparing tax returns and planning estates. Some paralegals coordinate the activities of other law office employees and maintain financial office records. Various additional tasks may differ, depending on the employer.

Paralegals are found in all types of organizations, but most are employed by law firms, corporate legal departments and various government offices. In these organizations, they can work in many different areas of the law, including litigation, personal injury, corporate law, criminal law, employee benefits, intellectual property, labour law, bankruptcy, immigration, family law and real estate. As the law has become more complex, paralegals have responded by becoming more specialized. Within specialties, functions often are broken down further so that paralegals may deal with a specific area. For example, paralegals specializing in labour law may concentrate exclusively on employee benefits.

The duties of paralegals also differ widely with the type of organization in which they are employed. Paralegals who work for corporations often assist attorneys with employee contracts, shareholder agreements, stock-option plans and employee benefit plans. They also may help prepare and file annual financial reports, maintain corporate minutes' record resolutions and prepare forms to secure loans for the corporation. Paralegals often monitor and review government regulations to ensure that the corporation is aware of new requirements and is operating within the law. Increasingly, experienced paralegals are assuming additional supervisory responsibilities such as overseeing team projects and serving as a communications link between the team and the corporation.

The duties of paralegals who work in the public sector usually vary within each agency. In general, paralegals analyze legal material for internal use, maintain reference files, conduct research for attorneys and collect and analyze evidence for agency hearings. They may prepare informative or explanatory material on laws, agency regulations and agency policy for general use by the agency and the public. Paralegals employed in community legal-service projects help the poor, the aged and others who are in need of legal assistance. They file forms, conduct research, prepare documents and, when authorized by law, may represent clients at administrative hearings.

Paralegals in small and medium-size law firms usually perform a variety of duties that require a general knowledge of the law. For example, they may research judicial decisions on improper police arrests or help prepare a mortgage contract. Paralegals employed by large law firms, government agencies and corporations, however, are more likely to specialize in one aspect of the law.

Familiarity with computers use and technical knowledge have become essential to paralegal work. Computer software packages and the Internet are used to search legal literature stored in computer databases and on CD-ROM. In litigation involving many supporting documents, paralegals usually use computer databases to retrieve, organize and index various materials. Imaging software allows paralegals to scan documents directly into a database, while billing programmes help them to track hours billed to clients. Computer software packages also are used to perform tax computations and explore the consequences of various tax strategies for clients.

Working Conditions: Paralegals employed by corporations and government usually work a standard 40-hour week. Although most paralegals work year round, some are temporarily employed during busy times of the year and then are released when the workload diminishes. Paralegals who work for law firms sometimes work very long hours when they are under pressure to meet deadlines. Some law firms reward such loyalty with bonuses and additional time off.

These workers handle many routine assignments, particularly when they are inexperienced. As they gain experience, paralegals usually assume more varied tasks with additional responsibility. Paralegals do most of their work at desks in offices and law libraries. Occasionally, they travel to gather information and perform other duties.

Training, Other Qualifications and Advancement: There are several ways to become a paralegal. The most common is through a community college paralegal programme that leads to an associate's degree. The other common method of entry, mainly for those who already have a college degree, is through a programme that leads to a certification in paralegal studies. A small number of schools also offer bachelor's and master's degrees in paralegal studies. Some employers train paralegals on the job, hiring college graduates with no legal experience or promoting experienced legal secretaries. Other entrants have experience in a technical field that is useful to law firms, such as a background in tax preparation for tax and estate practice or in criminal justice, nursing, or health administration for personal injury practice.

An estimated 1,000 colleges and universities, law schools and proprietary schools offer formal paralegal training programmes. Approximately 260 paralegal programmes are approved by the American Bar Association (ABA). Although many programmes do not require such approval, graduation from an ABA-approved programme can enhance one's employment opportunities. The requirements for admission to these programmes vary. Some require certain college courses or a bachelor's degree, others accept high school graduates or those with legal experience and a few schools require standardized tests and personal interviews.

Paralegal programmes include 2-year associate degree's programmes, 4-year bachelor's degree programmes and certificate programmes that can take only a few months to complete. Most certificate programmes provide intensive and, in some cases, specialized paralegal training for individuals who already hold college degrees, while associate's and bachelor's degree programmes usually combine paralegal training with courses in other academic subjects. The quality of paralegal training programmes varies; the better programmes usually include job placement services. Programmes generally offer courses introducing students to the legal applications of computers, including how to perform legal research on the Internet. Many paralegal training programmes also offer an internship in which students gain practical experience by working for several months in a private law firm, the office of a public defender or attorney general, a bank, a corporate legal department, a legal aid organization, or a government agency. Experience gained in internships is an asset when one is seeking a job after graduation. Prospective students should examine the experiences of recent graduates before enrolling in a paralegal programme.

Although most employers do not require certification, earning a voluntary certificate from a professional society may offer advantages in the labour market. The National Association of Legal Assistants (NALA), for example, has established standards for certification requiring various combinations of education and experience. Paralegals who meet these standards are eligible to take a 2-day examination, given three times each year at several regional testing centres. Those who pass this examination may use the Certified Legal Assistant (CLA) designation. The NALA also offers an advanced paralegal certification for those who want to specialize in other areas of the law. In addition, the Paralegal Advanced Competency Exam, administered through the National Federation of Paralegal Associations, offers professional recognition to paralegals with a bachelor's degree and at least 2 years of experience. Those who pass this examination may use the Registered Paralegal (RP) designation.

Paralegals must be able to document and present their findings and opinions to their supervising attorney. They need to understand legal terminology and have good research and investigative skills. Familiarity with the operation and applications of computers in legal research and litigation support also is important. Paralegals should stay informed of new developments in the laws that affect their area of practice. Participation in continuing legal education seminars allows paralegals to maintain and expand their knowledge of the law.

Because paralegals frequently deal with the public, they should be courteous and uphold the ethical standards of the legal profession. The National Association of Legal Assistants, the National Federation of Paralegal Associations and a few States have established ethical guidelines for paralegals to follow.

Paralegals usually are given more responsibilities and require less supervision as they gain work experience. Experienced paralegals who work in large law firms, corporate legal departments, or government agencies may supervise and delegate assignments to other paralegals and clerical staff. Advancement opportunities also include promotion to managerial and other law-related positions within the firm or corporate legal department. However, some paralegals find it easier to move to another law firm when seeking increased responsibility or advancement.

Employment: Paralegals and legal assistants held about 224,000 jobs in 2004. Private law firms employed 7 out of 10 paralegals and legal assistants; most of the remainder worked for corporate legal departments and various levels of government. Within the Federal Government, the US Department of Justice is the largest employer, followed by the Social Security Administration and the US Department of the Treasury. A small number of paralegals own their own businesses and work as freelance legal assistants, contracting their services to attorneys or corporate legal departments.

Job Outlook: Employment for paralegals and legal assistants is projected to grow much faster than average for all occupations through 2014. Employers are trying to reduce costs and increase the availability and efficiency of legal services by hiring paralegals to perform tasks formerly carried out by lawyers. Besides new jobs created by employment growth, additional job openings will arise as people leave the occupation. Despite projections of rapid employment growth, competition for jobs should continue as many people seek to go into this profession; however, experienced, formally trained paralegals should have the best employment opportunities.

Private law firms will continue to be the largest employers of paralegals, but a growing array of other organizations, such as corporate legal departments, insurance companies, real estate and title insurance firms and banks hire paralegals. Corporations in particular are boosting their in-house legal departments to cut costs. Demand for paralegals also is expected to grow as an expanding population increasingly requires legal services, especially in areas such as intellectual property, health care, international law, elder issues, criminal law and environmental law. Paralegals who specialize in areas such as real estate, bankruptcy, medical malpractice and product liability should have ample employment opportunities. The growth of prepaid legal plans also should contribute to the demand for legal services. Paralegal employment is expected to increase as organizations presently employing paralegals assign them a growing range of tasks and as paralegals are increasingly employed in small and medium-size establishments. A growing number of experienced paralegals are expected to establish their own businesses.

Job opportunities for paralegals will expand in the public sector as well. Community legal-service programmes, which provide assistance to the poor, elderly, minorities and middle-income families, will employ additional paralegals to minimize expenses and serve the most people. Federal, State and local government agencies, consumer organizations and the courts also should continue to hire paralegals in increasing numbers.

To a limited extent, paralegal jobs are affected by the business cycle. During recessions, demand declines for some discretionary legal services, such as planning estates, drafting wills and handling real estate transactions. Corporations are less inclined to initiate certain types of litigation when falling sales and profits lead to fiscal belt tightening. As a result, full-time paralegals employed in offices adversely affected by a recession may be laid off or have their work hours reduced. However, during recessions, corporations and individuals are more likely to face other problems that require legal assistance, such as bankruptcies, foreclosures and divorces. Paralegals, who provide many of the same legal services as lawyers at a lower cost, tend to fare relatively better in difficult economic conditions.

Earnings: Earnings of paralegals and legal assistants vary greatly. Salaries depend on education, training, experience, the type and size of employer and the geographic location of the job. In general, paralegals who work for large law firms or in large metropolitan areas earn more than those who work for smaller firms or in less populated regions. In addition to earning a salary, many paralegals receive bonuses. In May 2004, full-time wage and salary paralegals and legal assistants had median annual earnings, including bonuses, of $39,130. The middle

50 percent earned between $31,040 and $49,950. The top 10 percent earned more than $61,390, while the bottom 10 percent earned less than $25,360. Median annual earnings in the industries employing the largest numbers of paralegals in May 2004 were as follows:

Federal Government	$59,370
Local government	38,260
Legal services	37,870
State government	34,910

Legal Assistants, also known as paralegals, are assistants to lawyers. They interview clients, conduct legal research and draft legal documents under the supervision of lawyers.

Paralegal's or Legal Assistants help lawyers prepare for closings, hearings, trials and corporate meetings. They investigate the facts of cases and ensure that all relevant information is considered. They also identify appropriate laws, judicial decisions, legal articles and other materials that are important to cases. After they analyze and organize the information, paralegals may prepare reports that attorneys use in determining how cases should be handled. If attorneys file lawsuits on behalf of clients, paralegals help prepare the legal arguments, draft pleadings and motions to be filed with the court, secure affidavits and help attorneys during trials. Paralegals also organize and keep track files of all important case documents and make them available to attorneys.

Opportunities: In addition to private law firms, other organizations in the private sector hire legal assistants. Some sites include corporate legal departments, insurance companies, estate and trust departments of banks, hospitals and other health care organizations, real estate and title insurance companies and professional trade associations. Job opportunities in the public sector are available in community legal services programmes, consumer organizations, offices of public defenders, prosecutors and attorneys general, city attorneys, state and federal government agencies and the judicial system.

A good source of information on job openings is the local or state paralegal association. Many paralegal associations maintain job data banks or referral services and can provide you with a listing of firms and agencies in which legal assistant positions are available.

Career Advice for the Legal Industry

Temp in the Legal Industry

Our increasingly litigious society is carving a niche for temporary legal professionals, from paralegals and other legal support staff to attorneys, to help busy law offices cope with mounting caseloads. Do you have the skills needed for temping in the legal industry?

What it Takes?

Who are the most attractive candidates? "The bulk of all the temporary positions we handle involve document review, whether it be antitrust or litigation, so it's important to have experience with this in a large firm or corporate legal department," says Andrew Jewel, vice president of national operations for Hudson Legal. "Additionally, if you are familiar with electronic discovery, you should highlight the particular e-discovery review tools you have used."

Be sure to consider the following factors when evaluating your fit as a legal temp:

- Temporary legal staffers should have experience working in a private attorney's office, law firm or corporate legal department. However, recent law school grads might benefit from a temporary position as a paralegal while waiting to take the bar exam.
- Up-to-date computer skills are a must. Paralegals and attorneys will be more easily placed if they are versed in the use of Westlaw, the most common legal research tool.
- The ability to adapt to new personalities and disparate work environments will ease the way between temporary assignments.

Sign Up with an Agency

If your skills and experience fit the above criteria and you feel ready for the challenge, the next step is locating an appropriate temporary staffing agency. Cynthia Dale, who has worked as a temporary legal secretary in London, Los Angeles and New York, stresses the importance of having a strong relationship with an agency. "You need an agency that is willing to listen and when you find such an agency, you must be very clear about your own skills and expertise," she says. "Never exaggerate your abilities, because it will come back to haunt you. Instead, get yourself up to speed in the areas that will most likely help you get hired."

And be prepared for some rigorous pre-screening before you even have an opportunity to interview with the agency. You are probably going to be questioned in detail about your specific work experience and reasons for seeking legal temp work. Be prepared to sell yourself to the agency just as you would to a prospective employer. Moreover, the more detailed you are about your abilities and goals, the easier it will be for you and the agency to work toward finding an appropriate position.

For example, if you're looking for a temporary attorney position, Jewel recommends highlighting where you're admitted to the bar and whether you're a member in good standing. "Including where you're admitted helps us expedite our background check process ," he says. Jewel also likes to see the number of years of experience as well as number of large projects worked on, all in an easy-to-read resume that includes bullet points instead of large paragraphs.

Also, be honest with the agency about any weaknesses you have in interviewing skills or job searching techniques. Many agencies will offer coaching in these areas.

The Most Lucrative Opportunities

You're apt to find more jobs available for legal secretaries than for other positions. While paralegals can often clear their desks in advance of a short leave or vacation, attorneys generally cannot properly run the office for long without a skilful legal secretary.

Still, there are abundant temporary opportunities for paralegals and attorneys. Each is often brought in for assistance on specific projects or in the preparation stages for important trials or litigation.

According to Devry Shuart of Shuart & Associates, there is also ample opportunity to temp in the legal industry as a receptionist, an IT professional, a human resources professional, an accountant or as a court runner. "The jobs are definitely there, but you should still be sure to come in with a well-written resume formatted for emailing and be prepared, no matter which type of job, to prove your grammatical and computer skills," says Shuart.

Technology and Law a Growth Industry

The way we live changes continually as we reap the benefits of rapidly improving technology. Like technology, law also touches virtually every aspect of our lives and often interacts with technology. Technology makes television possible, but the law increasingly determines such things as the technical specifications of a broadcast license, the privacy rights of "newsworthy" individuals and the necessary compensation for creators of musical scores and screenplays.

Sometimes, as with photocopiers and computers, technology influences the practice of law. More often, however, it is the other way around. Also, in our complex society, lawyers and technologists must cooperate in addressing many important questions. Were the shreds of "pot" found in the car really tobacco? Did the axle break, causing the car to hit the tree, or did the axle break because the car hit a tree?

This article discusses the many legally-related job opportunities for people with technical training, whether they be engineers, physicians or scientists. These range from those who need only fairly basic legal knowledge, for example, to deal with contracts or regulatory processes to those who become full-fledged attorneys and practice patent and other intellectual property law.

Law-Related Opportunities for Technologists without Formal Legal Training: Problems involving both technology and law are common and growing in number and complexity. It may come as a surprise, but most of these are handled by people trained only in technology. Imagine, for example, two engineers negotiating a construction contract. Having agreed on specifications, dates and price, they will probably use a contract form drafted much earlier by an attorney neither of them even knows. Later, if one firm disputes whether the other is meeting its responsibilities, the grievance could be resolved by an engineer-arbitrator who, after a "trial," would determine the legal rights and responsibilities of the parties and possibly award money or compel action. Only if the losing party refused to comply with the arbitrator's decision, would lawyers be needed.

Or, picture two scientists dealing with a regulatory issue, for example whether a computer's shielding adequately prevents broadcast interference. One might represent a manufacturer; the other, the Federal Communications Commission. Again, attorneys would only be needed if they (or their supervisors) couldn't resolve matters without going to court.

Thus, many without law degrees do the type of work often done by attorneys. Whatever law they need is generally learned on the job or at short seminars and professional meetings. Also, longer Franklin Pierce Law Centre programmes may interest people who wish to know more about law without having to become a lawyer.

Patent Law as a Special Case: Owners of patents can prevent others from using new technologies for up to twenty years. The Patent and Trademark Office (PTO), an agency of the federal government, grants patents only after close scrutiny. Not everything new is patentable. It may not be in the public interest to give an inventor exclusive rights to some technology merely because he or she was the first one who needed it—particularly where it was available to anyone who cared to look. To obtain a patent, an inventor must first comply with the requirements of the PTO.

Since the 1790's, technically trained people have been helping inventors obtain patents. To seek patents on behalf of others, one must have the proper technical credentials and pass an examination offered by the PTO. Non-lawyers who are qualified to sit for (and pass) that examination become patent "agents." In an important 1963 case, the US Supreme Court recognized that patent agents "practice law" and that states are generally free to regulate such practice. Yet, it forbade states from interfering with the practice of non-lawyers admitted before the PTO.

Clearly, the most important service patent agents provide is drafting patent "claims." A patent's value is primarily determined by the scope and validity of those claims. Patents may cover the equivalent of a square inch of Arctic tundra or a square mile of Manhattan. If claims are too narrow, others can practice an invention without entering the territory reserved for the exclusive use of the patent holder. If they are too broad, covering what is already known or what is obvious to anyone skilled in the relevant technology, a court can find them invalid. Effective claim drafting and patent prosecution requires a unique combination of good technical understanding, oral and written communication skills and basic legal judgment.

However, patent agents are very restricted in what they can do. Unlike lawyers, agents cannot, for example, file appeals from the PTO to the courts, negotiate and draft licenses to use patented and unpatented technology, sue those who breach contracts or use software without licenses, or advise on ways to protect trade secrets. (Further, attorneys need no technical credentials to do anything other than prosecute patents before the PTO. Remarkably, as noted below, those involved in law and science or technology outside of the patent area, rarely do.)

In any event, many, if not most, patent agents eventually go to law school. It may be difficult for agents working without legal supervision to obtain malpractice insurance. Thus, most work under the supervision of lawyers. Consequently, agents are apt to resent that attorneys doing essentially the same thing enjoy higher status and income.

The Rewards of Being a Patent Lawyer: Patent law usually involves more than drafting and prosecuting claims—although it usually requires the capacity to evaluate their scope and validity. One lawyer, after practicing for 23 years in a private firm, explained the satisfaction of his profession:

I enjoy patent law because, almost by definition, I am working at the forefront of technology. I am fortunate in that my present clientele includes corporations and individuals with whom my engineering background is directly in point. My education in technology continues every day. Whereas before I learned the theory behind pertinent technology in academic surroundings, now I learn it from clients.

I have become increasingly aware of and involved in the business affairs of my clients. I especially enjoy licensing activity. Each license situation is unique and has its own technological and economic dynamics. In licensing, I can be innovative and creative to the benefit of my client. There is more of a people-aspect to licensing, with a different set of problems and a different measure of satisfaction for a successful result.

Engineers, physicians and scientists can do well in law school and be successful in any legal area that interests them. As a pilot-lawyer should have some edge in aviation law, a background in chemical engineering should be useful in something like environmental law. Yet, as mentioned above, other than for patent prosecution, nothing other than a law degree is required to practice in any area of law. In fact, lawyers filling environmental or other non-patent positions that involve technical issues are unlikely to have formal technical training. Hiring attorneys who lack such training rarely cede any advantage to those do and are often (sometimes very) put off by, *e.g.*, engineer-lawyers incapable of recognizing that technical issues may be only marginally relevant in disputes over the use of technology. So, the key consideration in finding a job in most law-technology areas is not whether you have a technical background but whether you have the skills and attitude hiring lawyers are looking for.

That aside, lawyers with technical training are scarce. Each year, about 40,000 law students graduate, but the total number of active patent attorneys, most of whom are technically trained, is perhaps 20,000. Patent law is a field with considerable demand. This translates into comparatively high salaries for those with good legal training and technical training relevant to important areas of technological growth. (Assuming the PTO would allow someone with a degree in buggy whip engineering, for example, to sit for the patent exam, passage would hardly generate a slew of lucrative job offers!) A 2003 survey of members of the American Intellectual Property Law Association found associates under the age of 30 to be making $95,000 to $144,000 (25th to 75th percentile) and senior lawyers, $600,000 or more (75th percentile, ages 55 to 60). Earlier, US News and World Report sketched an equally rosy picture, with beginning salaries of $70,000 to $90,000 and highs of $225,000 to $425,000.

Thus, many technically trained lawyers do intellectual property work because it is interesting, because they tend not to enjoy special advantages in other areas with both technical and legal components and because potential earnings are very attractive.

The Golden, Global Age of Intellectual Property Law: Intellectual property law includes perfecting, protecting and transferring patents, copyrights, trade secrets and trademarks.

Intellectual property rights are critical to continued investments in research and development and, hence, the competitiveness of US industry in domestic and world markets. As in the situation, for example, of those who invest in real estate development, or oil and gas exploration, if companies that discover new technology and bring it to market, often at great expense, could not profit from their efforts, they would soon be out of business. Thus, intellectual property rights are among the key tools of entrepreneurs and, in varying degrees, furnish an important foundation for the formation and growth of all companies.

This is said to be the information age. US News and World Report has likened ideas in the information age to iron ore and oil in the industrial age—essential raw materials. Because ideas cannot be literally fenced in or locked up, they must be protected by patents, trade secrets, trademarks or copyrights—the main forms of intellectual property.

Until recently, it was unclear that the fruits of biotechnology could be patented or that software could be protected by copyright. Even now, doubt remains about the extent to which software can be patented. Indeed, it has not been long since patents were generally regarded as the tools of "monopolists."

The past dozen years has encouraging technological innovation come to be regarded by both judges and members of Congress as important to preserving the US standard of living in the face of imbalances of international trade—and intellectual property has been seen as offering needed encouragement. For example, to secure uniform and reasonable recognition of patent rights, Congress recently gave one court exclusive jurisdiction over all US patents. More recently, to reduce the risk of illegal copying, it provided that software cannot be rented.

Intellectual property has become important not only here, but also in most other countries that increasingly recognize its role in improving the general standards of living. To encourage international harmony on intellectual property and trade, the US has recently changed our patent and copyright laws in fundamental ways.

Becoming a Lawyer: Full-time law school takes three years and culminates in the Juris Doctor. A J.D. from a school accredited by the American Bar Association qualifies a person to take the bar exam in any state.

As mentioned above, college graduates need not pursue any particular line of study to be accepted into law school. At Franklin Pierce Law Centre, for example, over a third of our students have degrees in engineering or science and many have extensive experience or advanced degrees, including MDs. and PhDs.—the last being particularly helpful for biotechnology patent careers.

LSATs and Grades: In law school admissions, an acceptable score on the Law School Admissions Test (LSAT) and good grades are important. Also, grades are usually evaluated in light of the difficulty of particular courses and the undergraduate schools' grading practices. For example, at least some admissions committees recognize that high grades are particularly difficult to get in engineering.

Specialty Programmes: Admissions personnel at schools without specialized programmes emphasize that patent or environmental lawyers must have a sound grasp of legal fundamentals. This is true, but it ignores several very important points.

- Core legal curricula throughout the US are very similar. Subjects such as legal research and writing, civil and criminal procedure, contracts, property and constitutional law are usually required.
- Of the 80 to 90 credit hours typically obtained in pursuit of the JD, a third or more will be elective. Few schools require many courses beyond the first year.
- Law school electives, of necessity, also cover fundamentals, but from differing perspectives. This could explain why a study of two graduating classes at a state law school found that enrollment in "fundamentals" courses did not correlate nearly as strongly with bar passage as one might expect. One course actually had a negative correlation!

Pay particular attention to schools with a good selection of electives in your area of special interest, but it is unwise to pay much attention to rankings. The alternative is to have to select many courses having little or no personal appeal.

Money: When comparing the costs of programmes, be sure to consider the cost of living—as well as income lost while in school and because of slow advancement. In light of such factors, tuition differentials may seem less significant, or disappear altogether.

Find Career Success in Court Reporting

The Legal Industry offers Variety and Growth: Options in the career field include: Judicial Reporting, Broadcast Captioning, Communication Access Realtime Translation (CART) Reporting and Realtime Reporting. The foundation for all these careers is capturing and converting the spoken word into information bases and readable formats.

The technology that reporters use is called realtime, which allows reporters to instantly convert their stenographic notes into readable English text. The text is then displayed on computer monitors or large projection screens. Realtime reporting is the only current method for actual voice-to-text translation.

Earning potential for the reporting professional is significant because there is high demand and the field includes opportunities for overtime compensation.

Average income for:

- Judicial Reporters, who are commonly known as official or freelance reporters and work in court or in depositions, is $62,000.
- Broadcast Captioners, who work for television stations or captioning companies, ranges from $35,000 to $75,000+, with opportunities for overtime depending on whether the captioner is on staff or works as an outside contractor.
- Cart Reporters, who supply communication access services for deaf and hard-of-hearing people, is between $35,000 and $65,000.

There are approximately 50,000 to 60,000 judicial, broadcast captioning, CART and Internet information reporters in the United States.

The US Department of Labour projects that reporting job opportunities will grow as fast as the average for all occupations through 2012. About 27% of the judicial reporters in the US actually work in court. Most of the rest are hires by attorneys to report depositions of potential trial witnesses.

Federal rules require captioning of hundreds of hours of live television programming each week, creating a surge in career opportunities for broadcast captioners.

To practice in the field, reporters earn either an associate's or bachelors's degree. In addition, they actively participate in continuing education programmes and pass stringent certification test of skill and knowledge, nationally or locally.

Judicial Reporters are at the centre of the most controversial cases-preserving history in war crime trials, the Scott Peterson trial, the Michael Jackson case, the Microsoft antitrust trial and hundreds of other trials and hearing every day- providing transcripts not only to the judge and counsel but also to the international media in hard-copy format or by posting to the internet.

Broadcast Captioning: Because of their skill with stenotype machines and computers, specially trained realtime reporters are in huge demand to provide captioning for live television broadcast. This service is called realtime captioning and primarily benefits the 28 million Americans who are deaf or hard-of-hearing by allowing them to follow news broadcasts, sporting events and other television programming. This service is especially valuable in an emergency; for example, during tornado or hurricane warnings, or during the tragic events of September 11th.

Cart Reporters: A version of the captioning process allows reporters to provide more personalized services for the deaf and hard-of-hearing people through Communications Access Realtime Translation (CART). Reporters work with hard-of-hearing students and people who are learning English as a second language, captioning high school and college classes as well as providing transcripts at the end of the sessions.

However, the classroom is only one environment where the skills of a CART reporter are put to use. CART reporters accompany deaf clients as needed-for example, to conventions, business meetings, doctor appointments, workshops, wherever communication access is needed-to provide an instant conversion of speech into text using the judicial reporter's stenotype machine linked to a laptop computer. And more and more, CART reporters are providing this critical service remotely, as an Internet or phone connection allows for immediate communication access regardless of location. Reporting agencies that specialize in this service cannot meet the demand.

National Court Reporters Association: The National Court Reporters Association has underwritten a short programme about careers in court reporting and captioning.

Law School for Mature Students

Do any of the following scenarios sound familiar? You have been working for years in a high-level career that is not taking you where you want to go...or you have been slaving away working on your Ph.D. and are facing many more years of fairly thankless work in other peoples' labs at student wages before you see any chance for autonomy or recognition...or you have one of those degrees (such as a degree in electrical engineering, computer science, or physics) you have heard all the intellectual property law firms are seeking. If you are a sophisticated, bright individual who has done a little poking around, you may have come to the conclusion that obtaining your law degree will be the solution to all your problems-your ticket to nirvana, the inside track to wealth and prestige! And it might even lead to interesting work.

Before you start filling out law school applications, it is critical that you understand the legal lifestyle and the proper strategy for crafting a successful legal career. Simply earning a J.D. is not enough. Depending on your ultimate goal, making a thoughtful transition into law requires your consideration of a number of factors before you take the leap.

This article will examine what an individual with an advanced degree or an established career needs to consider when approaching a transition into law. A subsequent article will examine how to have a successful career as a law firm associate if you come to a firm with an advanced degree (or degrees) or significant career experience.

So you want to be a lawyer? How much do you know about what lawyers do? Do you have friends in the law? Have you been perusing law firm websites, reading bios and hearing about those entry-level salaries for associates at top firms? Don't get caught up in fantasy. Do your research. Talk to people who are doing the work you think you might enjoy. Take them out to lunch. Ask about hours. Ask about how they spend their days.

The potential salaries are seductive, but be realistic. Most of the associates earning top dollar at the big firms are billing between 2,000 and 2,300 hours per year. And that doesn't include "non-billable" hours. This can be a shock to someone over thirty who is trying to balance family needs with the demands of being a junior associate. Many firms expect you to work into the evenings and on weekends. Many firms expect you to take on rush projects with very little notice. Many associates will say, "Your life is not your own." All of this is palatable if you are doing something stimulating that you enjoy, but it can be burdensome if you are doing work just to generate a paycheck. The firms paying the top salaries will also expect you to possess top credentials in terms of your law school and your grades.

Some people choose law as a second career for philosophical reasons. They want to become advocates for people, animals, or causes. Many of these individuals choose to work for the government, non-profit, or smaller firms where profits per partner may not be the sole criteria for judging success. If the zeal of advocacy, rather than the vision of making partner, is your driving motivation, you can take a more flexible route through law school.

Some choose law as a second career based on fascination or experience with specific subject matter. Many experienced individuals have had exposure to specific areas of the law in their professional lives, such as employment law, landlord rights, or real estate and can further their careers by getting the J.D. Again, the route through law school can be more flexible for these individuals, depending on the sizes of the firms they hope to join.

A J.D. can open the door to a multitude of opportunities within firms, within the corporate environment, within the non-profit sector, within government and for individual entrepreneurs. BCG specializes in placing attorneys in law firms; thus, these recommendations will focus on the individual pursuing a J.D. with employment in a law firm as his or her immediate goal. Before you consider where to apply to law school, it is imperative to know what type of law firm you may want to set your sights on.

Below are run-downs of some of the types of law firms you may be interested in joining.

Large, Prestigious National and International Law Firms: The top-tier national and international law firms tend to be large, with multiple offices, high-profile clients and significant profits per partner. These firms offer access to the premier lawyers in their practice areas, sophisticated and challenging work, the prestige connected with association with a top firm and opportunities to develop careers that may lead to national and international recognition.

These are the firms that pay the top associate salaries. However, most of these firms are very rigid in their hiring criteria. Many will only consider candidates from the top 20 law schools, as ranked by US News and World Report. Many also require GPAs in the top 10- to 20-percent range, law review experience and, in some cases, judicial clerkship experience. Competition to get job offers for beginning associates at these firms is intense. And you need to understand the process involved.

These firms hire the majority of their associates through their summer programmes. The firms conduct on-campus interviews during students' first and second years of law school to consider who they might invite to work for them as summer associates. If you have a successful on-campus interview, you will be invited to the firm for one or more call-back interviews. If you do well at these interviews, you may receive an "offer" to be a summer associate. Depending on how your experience goes during the summer, the firm may extend an offer to join them after you graduate.

If you set your sights on a prestigious firm, you need to have this process in mind when choosing your law school, approaching your studies and exams and getting involved in extra law school activities. Don't expect to be granted special notice or credit for life experience in law school. If you want to get into a top firm, your GPA will be a critical consideration.

Boutique Law Firms

Many individuals with concentrated experience in specific subject areas will set their sights on law firm boutiques that specialize in areas of the law that complement their backgrounds. A few examples include boutiques specializing in healthcare law, employment law, real estate and environmental law, municipal law, or intellectual property. Often, these boutiques may be more locally or regionally based. Attending a local law school with a decent reputation (ranked in the top 100) and having a strong background in the relevant subject matter may be enough to land you a first job. Grades will still be important, but hiring partners will probably take a broader view of what you are bringing to the table and what subjects you emphasized during law school.

If you know what area of the law you want to practice in, it may be smart to consider which law schools are highly rated for programmes specific to your area of interest. For example, in some cases, attending a school that is ranked highly in environmental law may be more important than attending a school in the top 20 that may not have a strong environmental programme.

However, be advised that the national boutique law firms that carry prestige and serve the big-name clients will probably be as demanding as the big general practice national and international firms. These high-profile boutiques want to be able to flaunt top-level credentials on their attorney bio-webpages and top-notch writing skills are a prerequisite.

Smaller Regional and Local Firms: Firms with 50 attorneys or fewer are incredibly variable in terms of their practice mixes and the types of junior attorneys they may want to hire. Some smaller firms are just as elite and particular as the large international firms. A good rule of thumb is to explore a firm's website and look at the attorneys' backgrounds and academic pedigrees.

The firm will be looking for junior associates with the same types of backgrounds. If you find a small firm that emphasizes the practice that interests you, make contact with one of its attorneys and find out if and how they bring in new attorneys. Most attorneys enjoy speaking with people who want to learn about what they do, as long as they are not asking for anything more than information. The process of hiring junior associates in these firms can be informal, based on word of mouth referrals or Internet advertising, or it can be based on a summer associate programme as rigid and formal as those at the larger firms.

The Special Case of Intellectual Property: Those individuals with strong science or engineering backgrounds make up a unique group. In some markets, intellectual property law firms are aggressively recruiting junior associates from among first-year law students who claim these credentials. If you have an electrical engineering degree, a computer science degree, a physics degree or a Ph.D. in the bio arts, as well as significant industry experience, a recruiter may be able to represent you as a new graduate.

The top intellectual property firms are selective and expect students from top-20 law schools, although students from law schools in the top 50 with strong intellectual property programmes may be considered. Grades are important, but extracurricular activities related to intellectual property are also weighed heavily. Speaking to a BCG recruiter with expertise in the intellectual property market while charting your course through law school will keep your options open, as you plan submissions to firms and progress through the associate ranks. A legal recruiter can be incredibly helpful long before you need him or her to help you find a job.

Conclusion: Do your research. Have some idea of where you are headed before you jump into law school. As a first-year law student, I had a sudden awakening. I thought I was taking time off from life to contemplate a new career, try things out, enjoy the luxury of simply being a student and explore whatever might interest me. I was jolted out of this reverie by the law school career counsellor who gave the first-year students their October orientation. She stressed that we were not allowed to submit resumes to law firms in pursuit of summer associate jobs until December of our first year. I seemed to be the only student surprised that we were expected to start approaching potential employers only 12 weeks after beginning law school. I remember walking out of that meeting dumbfounded, thinking about how I had no idea what I wanted to do after law school. How was I to know where to submit resumes?

Although not knowing what you want to do is not a terminal condition in your first year, if you end up setting your sights on the top-level firms, you will be behind in the game if you did not enter law school knowing that was your goal from the beginning.

Other considerations for the older student include the debt load you are willing to shoulder to achieve your goals in the law, the issue of geographic relocation and how well you will be able to balance law school and the legal lifestyle with your current lifestyle. I will explore these topics, as well as day-to-day strategies for "older" associates in sophisticated law firms, in future articles.

Changing Structure of Legal Services

Very rarely do academics get an opportunity to study both the internal operations of firms and the interaction of firms in an industry. Most empirical research follows either a panel of firms over time or the careers of individuals in a single firm. However, the Martindale-Hubbell Law Directory ("Martindale") contains detailed biographical information at both the level of the law firm and the individual lawyer for nearly all law firms and lawyers in America. Having data on firms and individuals across the entire industry allows us to link changes in the organizational structures and policies of individual firms to changes at the aggregate industry level.

In this study of the legal market, we use the Martindale data from 1998 through 2004 to explore, document and refute some "stylized facts" about the legal services industry. Because the Martindale directory contains a near census of the legal services industry, we have the opportunity to determine exactly what changes have occurred in the industry and whether a number of popular beliefs are true. Specifically, we will explore how changes in the nature of the profession may have implications for the structure of the legal services industry, the organization of law firms and the progression of legal careers.

By the "legal services" industry, we mean the group of private practice law firms that serve the legal needs of individual and especially corporate clients. While this industry is often referred to as "the legal profession," we avoid the use of this term because we wish to focus on those aspects of the profession-the organization and interaction of firms, the structure of these firms and the labour market that supplies them-that reflect economic and competitive forces rather than issues of professional norms and values. In this sense, this study treats the legal profession like a group of companies competing in an industry, trying to succeed by meeting the needs of its customers and employees.

Perhaps the most significant change in the legal services industry in the last twenty-five years has been the rise of corporate in-house counsels and the accompanying change in the nature of the relationship between law firms and their clients. This transformation, which some have described as "the decline of relationship lawyering," has led to a number of changes in the structure of firms, the structure of the industry and legal careers. Corporate clients have become less attached to their law firms and are more likely to "shop" for legal help. They have also brought more work in-house and changed the nature of work referred to outside law firms. The significance of "rainmaking," the process of attracting clients, has increased, leading to higher mobility in the legal labour market and an increase in meritbased partnership pay.

The consequences of these trends are many and varied. Some commentators have predicted the demise of the midsized, full-service law firm, arguing that this organizational form is likely to lose out in competition with larger firms and smaller, more focused boutiques who tailor the scale and scope of their services to meet the new needs of corporate counsel. Firms may also increase their geographic scope as an additional means of creating value for clients. Another consequence of the changing nature of client-firm relationships is a change in the structure of law firms. With partners spending more time on rainmaking, higher leverage is necessary in order to have enough lawyers to actually do the legal work. Along with this change in structure comes a change in the nature of legal careers: lengthening promotion "clocks" and the weakening of "up-or-out" promotion systems are both likely consequences of the increased reliance on business generation.

In this Article, we explore and document these changes in the industry, firm structure and careers. We find some, though not very compelling, evidence for a decline in midsized firms relative to smaller firms. We find a strong trend toward larger firms and especially large increases in the number of multi-office firms. We document increasing leverage in the industry, though this is mostly evident in smaller and midsized firms. We find that the average time to promotion is increasing across the industry and we find some provocative evidence for a decline in the use of up-or-out promotion systems.

The Martindale data is extremely rich and has the potential to address many interesting questions beyond the scope of the current analysis. In future work, we hope to explain the variation in the mechanism by which firms grow and shrink. In addition, we plan to study differences in firms' choices of human resource policies, including those that govern promotion, recruiting and training. Having data that tracks both firms and lawyers over time also provides a unique context for addressing questions about legal careers. In particular, we would like to consider how the progression of careers differs with respect to gender, lateral movement and specialization. Outside of the opportunity to study the legal services industry specifically, we can also use Martindale to shed light on other issues pertaining more generally to professional service firms.

The Article proceeds as follows: Part I documents the development of the data, Part II outlines the methodological approach, Part III presents our empirical results and Part IV discusses potential extensions.

Description of Martindale-hubbell Data

For our analysis, we are using data from the Martindale-Hubbell Law Directory. The Martindale directory has been published (in print form) for over one hundred years and is widely known as the legal services industry's most comprehensive professional directory. While there is no requirement that firms list, listing in Martindale has become standard practice for virtually all law firms in America. Martindale offers free basic listings for firms and lawyers, offers premium listings for a fee and charges subscribers for its directories. While sole practitioners and very small firms may not all list, we are confident that the Martindale directory represents a near census of the legal services industry for lawyers in firms of more than a few people.

Primary Data Download

We received quarterly data directly from Martindale starting from the first quarter of 1997, the first year that Martindale retained a historical copy of the directory in electronic format, through 2004. Because Martindale could not recover every historical quarter, there are a few missing quarters. The table lists all quarter-year combinations for which we have data. Martindale provided us with the following information on lawyers: name, law firm, office location, title, date of birth, date of admission to the bar, location of admission, law school name and graduation date, undergraduate school name and graduation date, practice areas and peer rating. We have the following information on firms: name, establishment year, office locations, main office and practice areas. For firms with premium listings, we also have biographies, profiles and representative clients. Martindale assigns every lawyer and firm a unique permanent identification number ("ID"). Within each firm ID, offices are assigned a unique ID number.

Cleaning Up the Data

1. The Consistency of Firm, Lawyer and Office IDs Since the Martindale dataset was gathered to be used as a directory, its primary structure is a cross-section rather than a panel. In order to use the data as a panel, we check for and correct cases where IDs are not longitudinally consistent. Our procedure for doing so is detailed in the Appendix.
2. *Elimination of Duplicates:* Lawyers sometimes appear in the database more than once in a quarter. We eliminate these duplicates (using a procedure documented in the Appendix) so that every lawyer appears at most once per quarter. Missing quarterly entries are discussed below.
3. *Assigning Lawyers to Positions:* The title information provided by Martindale separates lawyers across twenty-four different positions. Only a small fraction of firms

have no information on positions. Using information on the age distributions by position, we standardize Martindale's titles into five distinct categories: partners, associates, off-track attorneys, contract attorneys and retired attorneys.

4. *Aggregating Quarterly Data into Annual Data:* For analyzing time trends, it is important to compare data from year to year at the same point in time because the entry and exit of lawyers varies by quarter. While it is possible to use only data from a single quarter for each year, we instead annualize over all quarters in a year so as not to leave out lawyers who may not list in Martindale in every quarter. The major caveat of this method is that if a lawyer exits the legal services industry before the last quarter in the year, we will assign that lawyer to the last known firm he or she belongs to and, therefore, year-end totals may include lawyers that have actually exited earlier in the year. At worst, we delay exit by a maximum of three quarters. We aggregate the data to end with the first quarter of each year because a first quarter year-end allows us to include all available data. Because we have only one quarter of data for the year ending with Q1 1998, we count Q1 1997 in 1998. Including Q1 1997 helps to account for any lawyers that may have appeared in the missing quarters (Q2, Q3 and Q4 1997) but do not appear in Q1 1998. Relative to all other years of data, we will at most retain lawyers (who really exit the industry) for an extra three months. We have at least three quarters of data in all years other than 1998.

For each year, we convert the quarterly records into a single annual observation. If a lawyer's firm and position information does not change during the year, then this task is trivial. When a lawyer's firm (or position) changes, we use that lawyer's next and last firm (position) to determine which firm (position) appears second in chronological order. In the event that a lawyer appears in more than one firm (position) in the next or last year, then we order the data chronologically and assign the lawyer to the second firm (position). At the firm level, we keep the last known main office and branch office addresses in each year.

Restrictions to the Sample

Based on our estimates, the Martindale dataset contains records on about 950,000 lawyers per year working in law firms, as sole practitioners and for government, academic and corporate organizations globally. However, we are most interested in a subset of these lawyers. In most of our analyses, we restrict the sample to US lawyers working in private practice firms of five or more lawyers.

1. *Elimination of International Lawyers:* Although Martindale maintains some data on lawyers globally, we restrict our analysis to the US legal market because we do not have reason to believe that Martindale contains a census of international lawyers. For the small fraction of lawyers listed in US and international offices in the same year, we must make an assumption about which country they actually work in. Because firms frequently list lawyers in multiple offices, we apply an algorithm similar to that used in Part I.C. to eliminate duplicate office listings. The simple intuition works as follows: if a firm lists all lawyers in the main office, then we assume a lawyer listed in two offices in different countries works in the nonmain office. Specifically, lawyers listed in one US and one foreign office work in the office that is not in their home country (the international office, if the firm's main office is located in the United States; the US office, if the firm's main office is located outside of the United States). Alternatively, some firms list all lawyers in every office and, thus, we assume lawyers listed in more than two offices work in the country where their firm is headquartered.
2. *Elimination of Small Firms:* We have chosen to eliminate the smallest firms from our analysis for two reasons. First, because firms and lawyers choose to appear in the Martindale directory, we are concerned that Martindale may not represent a complete census of small firms.

Smaller firms are more likely to have localized demand and depend less on referrals, possibly making even a free directory listing not worthwhile. Second, we are concerned that the strategies, policies and practices of law firms (which are in the end the focus of our interest) may differ between very small firms and larger firms. It may be that very small firms are more like sole practitioners than they are like law firms. By eliminating the tiniest firms and sole practitioners, we rule out any problems that might result from drawing conclusions about law firm practices with data from organizations that are fundamentally dissimilar. Specifically, we choose a minimum cutoff size of five or more lawyers. The number of lawyers in firms of all sizes and the number of lawyers in firms of size five or more over time. The firms we exclude from our analysis account for a constant fraction of lawyers, about 30%, in all years. This 30% is a large number and (because sole practitioners are less likely to list in Martindale) is almost surely an underestimate of the number of lawyers not included in our sample. It is thus important to recognize that our analysis is not really representative of the entire legal services industry, but rather that part of the market served by law firms of five or more lawyers.

Our final dataset contains 2.1 million lawyer-year records and 110,000 firm-year records. Unless otherwise stated, all empirical results from this point forward derive from lawyers working in private practice firms of at least five attorneys.

3. *Elimination of Duplicate Office Listings:* Before we can calculate office size, we must assign lawyers to a single office in each firm-year combination. It is not surprising that firms may choose to list their lawyers in multiple offices since the function of listing in Martindale is to advertise legal talent and listing in more than one place makes it easy to find lawyers. For example, some firms list all lawyers in the main office and other firms list their name partners in every office. Some of these lawyers

may actually split their time between offices, but because other duplicate listings may be advertisements, we assume lawyers can only be in residence in a single office. We first use a lawyer's own career history to resolve instances where duplication results because lawyers change offices.

If a lawyer is listed in two offices in a year, but only one of those offices in the following year, we assume next year's office is the correct office for the current year. Next, we apply an algorithm that uses information about how many offices list a lawyer to assign the remaining cases of duplication. If a firm lists all lawyers in every (or almost every) office, then knowing the selection of offices a lawyer is listed in provides no additional information for assigning a lawyer to a single office. In this case, we assign lawyers to the main office. However, if a firm has many offices, but a lawyer is listed in only two of those offices, we assume the firm follows a practice of listing all lawyers in the main office and assigns such dually listed lawyers to the branch office.

The assignment for cases where a lawyer is listed in both offices of a firm that has only two offices is slightly more complicated. If a firm has only two offices, we do not know whether lawyers are listed in both offices because everyone is always listed in the main office (in which case duplicate lawyers should be assigned to the branch office) or because everyone is always listed in every office (in which case duplicate lawyers should be assigned to the main office). We assign lawyers who are listed in both offices (consistent with the assignment of lawyers who are listed in every office above) to the main office with a few exceptions. We call any office where office size equals firm size the "equal" office. If only one office in a firm contains all lawyers, we assume lawyers listed in both offices work in the office that does not list every lawyer, or the "not equal" office. If a firm has only two offices and neither office size equals firm size, but one office lists all partners in the firm, we assume lawyers listed in both offices work in the office that does not list every partner. Although these steps may seem to add an extra layer of complication, in almost every case the main office is the "equal" office and the branch office is the "not equal" office, making this procedure identical to the basic method detailed above. The key difference is that we refine our procedure for two-office firms such that we assign a lawyer dually listed in one office containing all lawyers to the branch office instead of the main office.

4. *Elimination of Demotions:* There are a few instances in the dataset where lawyers appear to move from partner to associate in the same firm. We assume these are data errors and that firms do not demote partners-that is, lawyers do not move from the partner position to the associate position while working at the same firm. Generally, we assume a lawyer is an associate until the last year in which he or she appears as an associate. For example, if a lawyer holds the positions associate, associate, partner, associate and partner in chronological order, we change the position in year three from partner to associate. We do not assume the alternative, that a lawyer is a partner from the first point in time that he or she is listed as a partner, because we find that many instances of demotions are the product of firms that list all lawyers as partners in one or two years, but separate associates from partners in all other years. We do, however, allow for demotions when lawyers move across firms because different firms may have different partnership criteria.

Other Lawyer and Firm Characteristics

1. *Lawyer Characteristics:* Martindale records each lawyer's biographical characteristics, including birth year and graduation year, on a quarterly basis even though this information should not change with time. Because lawyers are sent versions of their current profiles to revise each year, we eliminate inconsistencies in birth year by assuming the information contained in the most recent directory is correct. If a lawyer reports more than one law school graduation year, we assume the graduation year

that makes the lawyer closest to age twenty-five at graduation is the correct one. We calculate age as the current year minus birth year and years out of school as the current year minus law school graduation year.

We only observe the occurrence of promotion for lawyers who get promoted between 1998 and 2004. Ideally, we would like to know how many years each lawyer worked at his or her current firm (and prior firms) before promotion. However, our data are left-censored, meaning that we have no information about the careers of lawyers prior to 1998. Hence, it is not possible to track a lawyer from the beginning of his or her career to the point of promotion. We can calculate years to promotion as the length of time between a lawyer's graduation from law school and year of promotion. This procedure is consistent with practice in many firms, who calculate a lawyer's "class" not by time at the firm, but by time out of school, including clerkships, as years that count on the partnership clock.

Martindale does not track gender for each lawyer, but we are still able to construct a measure of gender using frequencies of names by gender from the 1990 Census. From the Census data, we calculate for each name the percentage of people with that name who are female. If the percentage of females is greater than 50% for a name, then all lawyers with that name are coded as female. If the percentage of females is less than 50% for a name, then all lawyers with that name are coded as male. For example, 3.1% of females and 3.6% of males are named Kerry, thus, 46% of people named Kerry are female and all lawyers with the name Kerry are coded as male. The Census only lists names accounting for 90% of the population, so we use Baby Name Guesser, a web-based tool, to fill in the most likely gender for any remaining names. There are about 1% of lawyers for which we are unable to identify a gender, either because a lawyer reports only his or her initials or the name is so rare that it does not appear in either of our sources.

2. *Firm Characteristics:* We build firm statistics by aggregating up from the lawyer level. Since we know each lawyer's office and firm, we can easily compute office and firm size by counting the number of lawyers in each office and firm. A multi-office firm has more than one populated office location. If the office assignment procedure reduces the number of lawyers in an office to zero, we do not include it in the firm's total number of branch offices. We calculate leverage, a measure of a firm's hierarchy, as the ratio of associates to partners. We measure the size of a firm's geographic footprint as the smallest region that contains all of a firm's offices.

Descriptive Statistics

The total number of lawyers reported in the Martindale data over the time period starting in 1998 and ending in 2004. It also shows the number of lawyers working in firms of five lawyers or more and the number of these lawyers in the top ten legal markets. There is a slight upward trend in the total number of lawyers during this period.

Almost one-half of all lawyers working in firms of five or more work in the top ten legal markets (as ranked in 1998 by the number of lawyers working at firms headquartered in each metropolitan statistical area ("MSA")), but the top ten markets do not account for an increasingly greater fraction of lawyers over time. Nor has there been very much change in the relative size of the top ten legal markets. New York is the city with the most lawyers and Washington, D.C. is second with about half as many lawyers in our sample. Even though all markets are growing, no top ten city is gaining disproportionately: there has been no change in the ranking of market by size over this period.

Though the number of lawyers is growing, the number of firms is declining slightly.

Thus, the average size of firms is growing. That the number of law offices grew until 2000 and has been stable since then. Therefore, firms are growing by starting and especially acquiring, new offices. The addition of offices is accompanied by an increase in geographic diversification, showing the number and percentage of lawyers by the size of firm footprints. Firms with a national footprint account for 9% of lawyers in 2004, up from only 4% in 1998. In spite of this trend, the market is still largely local: over 45% of lawyers work at firms whose offices are all contained in a single MSA, with 20% more working in firms with a geographic reach no larger than a single Census subregion.

In 2004, the pool of lawyers is still predominantly male, with women increasing as a fraction of all lawyers from 23% to only 26% over the seven-year time period. However, the percentage of women among all lawyers is not representative of the current level of gender equality because the existing stock of lawyers is mostly male. We instead calculate the percentage of women among the flow of entrants into the profession. We find that women have made significant inroads, increasing from 40% of all first year associates in 1998 to 47% by 2004.

Methodological Approach

Since this is the first study of the legal services industry using the entire Martindale directory, our primary research approach is to let the data tell the story of the industry from within rather than imposing (and testing for) a specific hypothesis ex ante. In the results that follow in Part III below, we present a series of descriptive statistics that document changes in the industry that may have been brought on by the decline in relationship lawyering. Since we are essentially using a census, we have the opportunity to compare our results with conventional wisdom derived from anecdotes, theories, or small sample analysis.

An advantage of our data is that we are working with the entire population of lawyers and not drawing inferences from a sample. Ordinarily, empirical research focuses on finding statistically significant differences, attempting to show that a difference in the observed sample is indicative of a difference in the population. But, when we observe a change in a variable over time, we know that this change is the actual difference in the population. For this reason, we almost never report standard errors or statistical significance levels in this Article. Our burden is instead to identify whether and where there have been interesting or significant changes in the industry.

We describe each variable of interest using some combination of totals, means, medians and distributions. Means (and other measures of central tendency) can be misleading indicators when the underlying distribution is not symmetric, has more than one peak, or is changing over time. For this reason, we will often try to present results that show the entire distribution of the data, allowing for a fuller analysis of the underlying phenomena.

A serious shortcoming of our data is their short time span. While the Martindale directory has been published for over one hundred years in printed form, the company has only kept its historical files in (incomplete) database form since 1997. For this reason, we have data only beginning in 1998. Clearly, many of the changes that we examine in this Article occurred much longer ago than this and so many of the trends that we document have been going on for much longer as well.

To the extent that we discern changes in the structure of the industry, of firms, or of careers over this short time frame, we believe that it is even more powerful evidence of important changes occurring in the legal profession, over a longer time period. However, failure to find changes over this shorter time period must be viewed with skepticism.

Analyzing the Market for Legal Services

Effects of the Decline of Relationship Lawyering : In this Part, we analyze four trends in the legal industry: the changing distribution of firm sizes; the rise of multi-office, geographically-dispersed firms; increasing leverage in law firms; and changes in the nature of legal careers, particularly the lengthening of the partnership "clock" and a possible decline in the use of "up-orout" promotion systems.

We argue that all of these trends may be linked to the changing nature of the relationship between law firms and their clients.

Over the past quarter century, much has been made of the rise of the in-house counsel and the resulting demise of "relationship lawyering." Corporate legal departments have exhibited significant growth since the early 1980s and have continued this trend in recent years. Between 1998 and 2004, the 200 largest in-house legal departments grew from a total of 24,000 to 27,500 lawyers. Armed with more talent and the goal of cutting costs, corporate law departments are performing an increasing share of legal work in-house. General counsel surveyed in 1998 did not anticipate sending more work to outside law firms in the coming year, despite a belief that the volume of legal work would increase. Cost pressures are also leading corporate counsel to reduce the number of outside law firms they use.

One consequence of the increasing size and scope of in-house legal departments has been that corporate clients no longer hire a law firm to "be their lawyer," developing a multifaceted and lasting relationship and giving that firm virtually 100% of their legal work. Galanter and Palay document an increase in the size of corporate law departments because corporate clients now do much of their routine legal work inhouse, they are more likely to hire outside firms only for large, complex, or specialized matters. These firms are often hired for a specific transaction and corporate clients seek to hire the most capable lawyer for each case. "Increasingly, clients view their legal counsel as a commodity that can be purchased from a number of sources" and often "shop" their work through a competitive selection process to a broader set of firms. Long-standing relationships between clients and firms have been replaced by a transactional system where clients hire individual lawyers, not law firms.

In response to these changes, firms face two possible strategies to compete. As detailed below, they can either expand in terms of size and service offerings, or they can specialize. Lost in this dynamic is the traditional midsized, general practice firm-the demise of which has been predicted for years. In the analysis in Part III,B below, we look for this decline and find only modest evidence of this trend.

Another way that firms can compete with larger in-house legal departments is to increase their geographic diversity. It is unlikely that in-house corporate legal departments can build sufficient expertise in multiple jurisdictions. Thus, firms that can offer this type of expertise will be able to offer clients value. We document the increasing geographic diversity of law firms in Part III.C.

As old relationships become less important, law firms (and lawyers) now must compete against each other for clients in a way that would have been unimaginable in the past. Rather than relying on competent legal expertise and loyalty to the firm to maintain clients, partners must actively seek out new clients. Rainmaking is now a necessary tool for a law firm's performance and survival. "As a result, firms-more than ever before-are grappling with how to teach their associates and new partners to be good marketers as well as good lawyers."

The increased importance of business-getting changes several aspects of legal practice, legal careers and the structure of law firms. Since the senior members of the firm must spend more of their time getting business, they have less time to actually do the legal work. This implies that having more associates per partner may be an optimal response to this change in the nature of the legal profession. We document this trend in Part III.D.

There is another implication of the increased importance of rainmaking to law firms. Associates now must do more than simply show that they are "good lawyers"; they must show they are good business generators if they are to be promoted to partner. This may imply the need to lengthen the amount of time before an associate is elevated to partner.

In addition, firms finding that associates are more valuable as associates, but less valuable as partners, may reduce their use of up-or-out systems to keep associates at the firm for longer periods. In Part III.E, we examine changes in promotion rates and systems. We find evidence that promotion clocks are lengthening and mixed evidence of the reduced reliance on up-or-out career systems.

Firm Size

With the expansion of corporate legal departments, the type of work referred to outside law firms has changed and the degree of cost competition has increased. Corporate clients use law firms when they need capacity or additional expertise, as may be necessary in complicated, multidisciplinary, or specialized work. Given the increased competition for business, law firms are altering their scale and scope to best meet the demands of clients.

A growth strategy is one means of both solidifying existing client relationships and developing new ones. First, by increasing in size, law firms can meet the capacity needs of corporate clients. Second, by increasing scope, or adding complementary practice specialties, law firms can represent clients in transactional matters requiring legal expertise in several areas. Firms adding scale in every practice area can offer their clients "one-stop shopping," which may build ties to the firm lasting longer than a single legal matter through active cross-selling.

Purchasing all legal services from a single firm for a multidisciplinary matter may also be less costly than purchasing these services at carte from different law firms. Because these moves help to protect client bases, they also keep the most powerful rainmakers (and highest revenue generating partners) from looking for new firms with greater scale or scope.

However, firms need not necessarily get bigger in order to survive. With the demise of relationship lawyering, corporate counsel are inclined to hire the best firm for a specific and likely specialized matter. Boutique firms succeed because they offer highly-specialized and narrowly-focused expertise without the high overhead costs associated with many large firms.

While growth and the expansion-or alternatively reduction-of practice areas are consistent responses of law firms to the competition for clients and the increasingly specialized nature of legal work, firms that do not adapt risk failure. Whereas midsize general practice firms would have satisfied the routine legal needs of clients in the relationship model, they are hard pressed to compete against firms that offer more depth in each practice area in the transactional model. Midsized firms are also more likely to be victims of poaching and exit inertia:

With the proliferation of rapidly growing branch offices and the expansion of existing large firms, rainmakers have an everincreasing number of lucrative career alternatives. A rainmaker at a midsize firm who now labors under the knowledge that his or her compensation is substantially less than market value and that he or she is, in effect, losing hundreds of thousands of dollars by remaining with the firm, will be hard-pressed to stay. The loss of rainmakers has serious consequences beyond the loss of business itself.

The departure of high revenue generating partners will make it even more difficult for midsize firms to attract and maintain clients and lawyers, potentially leading to more departures and eventual dissolution.

In order to identify whether there has been a shake-out of midsized firms, we calculate the distribution of lawyers by firm size. We weight our analysis by lawyers because the largest 15% of firms account for about 60% of lawyers. A lawyer-weighted statistic measures the relative dominance of firm size categories in terms of their share of the legal labour market. Tracking the percentage of lawyers rather than firms also allows us to capture both the addition (or deletion) of firms to a size bucket and the continued growth (or decline) of the firms already in that bucket.

As a baseline, we create firm size deciles, setting the minimum and maximum firm size for a decile such that each decile includes roughly 10% of all lawyers in 1998. Only the first two deciles contain more than 10% of lawyers. The lumpiness in the distribution occurs in the smallest firm sizes because each decile contains relatively few different firm sizes and many firms per size. If we were to, for instance, make the smallest decile those firms with five lawyers (instead of five to six lawyers), this bucket would contain many fewer than 10% of all lawyers.

If it were true that midsized is no longer a sustainable firm size, we would observe growth over time in the tails of the distribution and a decline in the middle of the distribution such that the new distribution would be U-shaped relative to the 1998 distribution. A U-shaped distribution implies that more lawyers either work for large firms or for small firms in 2004 than in 1998. The distribution of lawyers by firm size in 2004 indicates that this is not the case. It is true that there has been substantial growth in the largest firms. Firms with 389 or more lawyers account for slightly less than 10% of lawyers in 1998, but over 18% of lawyers in 2004.

Thus, the largest firms are growing, perhaps in response to the need to offer clients the capacity and the collection of practice areas needed for complex, multidisciplinary legal work. We see no evidence, however, of a corresponding rise in small firms or a larger decline in midsized firms. In fact, it appears that the largest firms have grown at the expense of firms of all other sizes. The net change in the percentage of lawyers by firm size, or the difference between the distribution of lawyers in 2004 and 1998. From this viewpoint, it is again apparent that a higher percentage of lawyers work at the largest firms, but also that the losses at midsized firms do not account for relatively more of the decline.

Before we can rule out a decline in the midsized firm, we must reconsider our definition of midsized. The analysis above aggregates firms nationally and effectively assumes that similarly sized firms are the same, irrespective of their size, relative to other firms in their local markets. So, the largest firm in Boise, Idaho may be lumped together with a midsized firm in New York, potentially clouding aggregate results.

If the market for legal services is more local in nature or is at least more local for midsized firms, then an analysis by market allows the definition of a midsized firm to vary by location. We repeat the size distribution analysis limiting the data to the top ten legal markets. We use the same size deciles as before, but instead calculate the fraction of lawyers in the top ten markets working at firms in each decile.

Since size is a firm-level characteristic, we include all lawyers working at firms headquartered in the top ten legal markets even if the lawyer actually works outside of the top ten markets. We find some evidence that firms in the middle of the distribution are exiting (at least relative to the smallest firms), but there is still no obvious growth in smaller firms.

Even looking at only the top ten legal markets, the fact that we do not observe an increase in small firms could still be a result of aggregation. What counts as a small firm in Houston may differ from what is considered small in Boston. And certainly this is true for New York. We calculate the net change in the percentage of lawyers by size decile for each of the top ten legal markets.

We present only the results for New York and San Francisco. We find an even more pronounced decline of the midsized firm and now see some growth in smaller firms. This is especially true in San Francisco, where the boom in boutique firms serving start-ups and practicing intellectual property law may show up in the data as growth in most of the smallest five deciles. The results are similar when looking at the remaining top legal markets, though not all markets exhibit growth of smaller firms.

Geographic Diversification

In addition to increasing their practice area scope, firms may also increase their geographic scope (and regional expertise) as another means of filling a void in in-house capabilities. As business is increasingly more global (and certainly more national), having offices located across the country enables law firms to offer representation in geographic areas where corporate counsel are less familiar with local laws. A large regional presence also may help law firms maintain client relationships as corporate counsel push not only to hire the best firm for a transaction, but also to reduce excessive legal costs associated with hiring a different team of lawyers in every geographic locale.

The observed growth of the largest firms is not inconsistent with an expansion of geographic locations, since firms can grow both by adding lawyers within their existing network of offices or by extending their network (*e.g.*, by opening or acquiring new offices). Over this period, average firm size grew by 13%, while average office size grew by only 6%. Since office size is not increasing by as much as firm size, it must be the case that law firms are adding new offices to their network. An increase in the total number of law offices over time with a decrease in the number of law firms implies that the number of branches per firm must have increased on average.

However, the average (unweighted) number of branches per firm has only increased fractionally over the past seven years. This average is misleading because although multi-office firms account for just 20% of firms in 2004, almost 60% of lawyers work at multi-office firms. An average that weights the number of branches per firm by the number of lawyers who work at each firm instead represents the number of branches in the average lawyer's firm. Hence, the average number of branches per firm is 1.3 in 1998, but the average lawyer works at a firm with 2.8 branches.

We calculate the distribution of lawyers by the number of branch offices. In 1998, 50% of lawyers work in multi- office firms and over time the percentage of lawyers working at firms with at least five offices has increased. By 2004, 57% of lawyers work in multi-office firms and over 10% of lawyers work at firms with ten or more offices. More lawyers work for the law firms with the most offices, another indication that firms are adding scale in every geographic area in order to better compete for corporate clients.

The branch distribution analysis for firms based in the top ten legal markets. Firms with the most offices gain an even higher percentage of lawyers relative to firms with less than ten offices than compared to the analysis including all cities. For almost every possible number of branches per multi-office firm, a higher percentage of lawyers in the top ten cities work at those firms than compared with the same distribution for all cities. Thus, geographic expansion is especially prevalent for firms most likely to be serving the needs of global corporations.

Leverage

With the demise of long-standing client relationships, partners are spending an increasingly larger fraction of their time generating business. As a direct result, law firms need more lawyers to spend time doing legal work. Not wanting to dilute the rainmakers' earnings, which may result in the loss of business-finders, firms are likely to add business-minders at a non-partner level.

With the addition of associates and a constant number of partners, leverage will increase. However, the increased emphasis on rainmaking and number of competitors competing for the partnership "prize" reduces the likelihood that associates are promoted. A smaller expected prize may discourage associates from working long billable hours and, as a result, firms may have to increase leverage even more to counteract a reduction in hours per associate.

Increased leverage may not only be a consequence of increased rainmaking. A high leverage ratio ensures that firms have a supply of relatively cheap labour to meet the short-term capacity needs of large clients. Optimal leverage is also connected to the level of complexity and specific specialty of the legal work.

We calculate the average leverage ratio of all firms and firms based in the top ten legal markets. We exclude in our analysis of leverage those firms for which leverage is zero in every year. A zero value of leverage implies that a firm consists of all partners and no associates. This may either occur because a firm really has no associates or because the firm chooses to only list its partners in Martindale. Many smaller firms likely have no associates, but this is less plausible for larger firms. If a firm ever lists associates, we assume that in a year where that firm does not have associates, it is the case that they really do not have associates and not that they choose not to list their associates.

If a firm never has associates, we cannot differentiate between the two possible explanations and we err conservatively by omitting all firms in question from our analysis. The average leverage for all firms and firms based in the top ten legal markets move in tandem, but the average leverage for firms in the top ten cities is higher than the leverage for all firms. Median leverage is less than the mean leverage.

The value of average leverage seems surprisingly low. One explanation is that our definition of a partner is based on what firms report to Martindale, which likely includes both equity and nonequity partners. It would not be surprising if firms add lawyers to do legal work at the nonequity partner level. Nonequity partners have significant experience useful for representing clients in complicated legal matters, but do not dilute the earnings of the highest revenue earning equity partners.

According to the chairman of Duane Morris, "nonequity partners fill a very important role at a firm. You need business generators, but you also need people to do the work. To the extent that firms use nonequity partners as an additional form of leverage, leverage is higher than indicated in our analysis. In addition, if the use of nonequity partners is increasing over time, as suggested by Henderson, then we understate the increase in leverage that may be occurring. We have also omitted off-tracks, a title that was historically reserved for semiretired partners but is increasingly being given to permanent associates, from our leverage calculation. This omission also leads us to underestimate the level and any increasing trend in leverage.

We segment our analysis by size decile to test whether larger firms are indeed more leveraged and determine whether the increasing trend in leverage results directly from the increase in lawyers working at large firms. We present results for both the means and the medians since there are several large leverage outliers in the middle size deciles.

The average leverage in decile seven jumps in 2001 because a single highly-leveraged firm has two partners in 2000 and only one partner in 2001, thus doubling its leverage. Leverage is higher at larger firms, increasing almost monotonically across the size range. However, leverage appears to be increasing through time in the small- and medium- sized firms, but not in the largest firms. This pattern is not entirely consistent with the argument that the increasing demand on senior lawyers to bring in new clients is forcing firms to hire more junior-level lawyers. If this was the reason, we would expect to see at least as much increase in leverage at the large firms as at the small.

We also segment our analysis of leverage by firm types in order to separate law firms by the kinds of clients they might serve. Firms that have an office (including branch offices) in New York are more leveraged on average than firms without a New York office. Multi-office firms are also more leveraged on average than firms with only a single office. To some extent, this may be true because firms with a New York office or multiple offices are more likely to be larger and the largest firms are the most leveraged.

In order to disentangle the effects of size, location and structure from the time trend in leverage, we perform multiple regression analysis attempting to explain leverage ratios in our firms. We run the following regression:

The results of this analysis indicate that some of the results to hold, while others do not. There is a highly significant time trend of increasing leverage. Firms with New York offices are more leveraged, as are multi-office firms. However, the multiple regression results indicate that, after controlling for the effects of New York City and multiple offices, the smallest and the largest firms are more leveraged than are midsized firms. The leverage of firms in the eighth and ninth deciles is almost always lower than that in smaller firms.

Although the overall increase in leverage is broadly consistent with a trend towards greater division of labour between senior and junior lawyers and an accompanying need to hire more junior lawyers, some of the patterns (particularly looking across firms of different sizes) are not consistent with this story.

Legal Careers

As the importance of soliciting business increases for partners in law firms, current partners may find it more difficult to make new partner decisions. When clients were tied to firms by history and a network of relationships, a young lawyer could succeed in his or her career by simply serving the specific needs of one or a small number of clients: this was grounds enough for promotion. However, since clients are now more likely to shift firms, associates must prove that they can also develop business if they are to be promoted to partner. Law firms are reluctant to admit new partners unable to generate business because rainmakers are highly mobile and any dilution of partner earnings might cause important partners to depart. In response to this difficulty, firms may extend the promotion track to allow more time for associate evaluation and decrease the likelihood of making promotion mistakes that might dilute earnings.

Firms are effectively lengthening the promotion clock to full partner by creating permanent non-partner positions: through the addition of nonequity partnership tiers, reducing the use of up-or-out promotion policies, or creating off-track positions. Since the jobs of associate and partner are now more different, moves that aim to keep associates at the firm, in any capacity, are also consistent with the need for more lawyers to do actual legal work.

Two-tier partnerships enable firms to distinguish between promising newcomers and senior rainmakers. Firms can test out these prospective equity partners and reap the benefit of the thousands of hours per year that these ambitious young lawyers bill. Their advance to the more prestigious level of senior partnership will depend on their legal skills and their ability to generate business.

As we have already stated above, the Martindale data does not enable us to distinguish equity partners from nonequity ones, but the increase in the number of two-tier partnerships is well-documented. The Martindale data does allow us, however, to measure the timing of promotions and to examine whether firms are relying on up-or-out promotion systems.

That, despite an increasing number of associates, the number of promotions has not increased. Since the number of associates is increasing during this time period, the percentage of associates promoted must have decreased. A longer time to promotion is consistent with the fact that "associates need more time to hone rainmaking and practice skill." The average time to promotion is increasing across all size deciles, but is slightly longer at the smallest firms than the largest firms. Perhaps this results because lawyers who fail to make partner at the largest firms move laterally to smaller firms where they attempt to make partner again. These new firms likely evaluate associates' abilities in their own firm for several more years before making a partnership decision. Table suggests that partnership clocks are also affected by the firm's market: the average time to promotion is slowest in New York and the average time to promotion is the fastest in the high growth cities-Atlanta, Dallas and Houston.

If a firm follows a strict up-or-out policy, then there simply would not be any associates beyond the year in which associates are eligible for promotion. What a strict up-or-out policy might look like in the data. A softer up-or-out policy would look more like Figure, with some associates leaving the firm (or getting promoted) before the promotion year and some staying on as associates after the promotion year. Irrespective of associate exit patterns before promotion, a strict up-or-out policy implies that at the point in time where the firm evaluates associates (the "promotion year") they are either promoted or let go. Thus, the firm would have no associates of tenure greater than the promotion year. A weakening of the up-or-out system implies that firms retain associates whose peers have already been promoted. In short, existence of a tail to the right of the promotion year is evidence that the firm is not strictly following an up-or-out policy.

In order to measure the extent to which the industry follows an up-or-out rule, we plot the total number of associates by years out of school. The slow increase in the early years indicates that lawyers move into the industry (possibly from clerkships or other government jobs) up until about year four or five. Rather than a vertical drop, the number of associates declines slowly but steadily over time. The lengthy tail, well beyond the median promotion age, indicates that not all firms follow a strict up-or-out policy. To some extent the slope of the line, especially around the median promotion age measures the degree to which firms follow an up-or-out policy.

The steeper the slope, the more closely firms apply an up-or-out rule. The slope does not appear to have changed much over time, indicating that up-or-out is actually still the predominant practice. The horizontal shift from 1998 to 2004 is consistent with a lengthening of the promotion clock.

One problem with the analysis above is that it treats all firms as if they have the same promotion policies. But, in reality, the timing of the promotion decision (as well as the use of an up-or-out policy) varies by firm. Analysis of the sort performed in Figure is again confounded by aggregation problems. To see this, imagine that all firms used strict up-or-out policies, but some promoted at Year 9, some at Year 10, some at Year 11, etc., then it would appear (at the industry level) that up-or-out was weak or nonexistent.

A potentially better way of measuring up-or-out policies is to attempt to discern whether individual firms are using the policy. We do this in the following way. We calculate, for each firm in each year, the median years to promotion (the "promotion year"). We then ask what percentage of associates at least as tenured as the promotion year but not yet promoted in that year are still associates at the firm two years later. We aggregate this ratio across firms and look to see whether, through time, it appears that firms are less strict in their up or out policies. We find some evidence for the weakening of up-orout at the individual firm level. In 1998, on average, 44% of those not promoted by the promotion year are still at the firm as an associate two years later. By 2004, this percentage had risen to 50%. Thus, firms may be weakening up-or-out policies in response to the increased reliance on business generation for a successful partnership.

As an alternative to the permanent associate position some firms may instead weaken up-or-out with the use off-track positions. These firms are essentially retitling lawyers who are not promoted to other permanent non-partner positions. We calculate the number of off-track lawyers by years out of school in figure. There are more offtrack lawyers in 2004 than in 1998 across all years out of school and this holds in the range of years around which a lawyer might have been passed over for promotion. If we repeat this analysis at the firm level, we find in 1998 that, on average, 4% of associates at least as tenured as the median promotion age remain at the firm in off-track positions two years later and this percentage is constant through time. Thus, while we find some evidence that firms are using offtrack positions as an alternative to the up-or-out model, they are not actually increasingly using these titles over time for associates beyond the promotion age.

Future Research

The Martindale directory is a rich dataset, ripe with possibility. The analysis presented in this Article is merely suggestive of the kinds of questions that could be answered. Having developed a basic understanding of how law firms are structured both at the firm level and at the industry level will provide context for addressing a broader set of industry and organizational issues in the future. While we have emphasized that the largest firms are getting larger, we have not yet attempted to explain the process by which these firms are growing. Some firms may make acquisitions and others may grow organically. Not all firms are growing; some are clearly downsizing too. The Martindale data presents us with the opportunity to study both how firms expand and shrink. We would like to establish what strategies firms follow and address how variation in growth strategies affects their choice of human resources ("HR") policies.

In addition to the impact of growth on HR policies, we can also study more generally why firms vary in their choice of HR programmes. Some have argued that firms invest more in young associates than they recoup in return. Why then do some firms choose to hire straight out of law school (and train their lawyers) and others do not? Why do some firms hire lawyers only from the elite law schools while other firms hire from a broad array of schools? Why do firms vary in the length of their associate tenures? In the strictness of their up-orout policies?

There are also a series of interesting questions to ask related to legal careers and how lawyers' career paths differ. We can identify how gender affects the career progression of associates and partners. And, we can determine what kinds of firms do more to promote gender equality. Lateral movement is also thought to be much more common than it used to be, even at the associate level. Again, we can examine how the career paths of lateral movers differ from those who never move and compare the careers of lawyers who move as individuals to those who defect in groups. We can also ask whether the career progression of lawyers who are narrowly specialized differs from those with a general practice area and whether the point in time at which a lawyer begins to specialize matters for promotion.

Without data on economic performance, it is difficult for us to make normative statements about the success or failure of particular firm strategies. We can look at organizational growth or longevity as signals of success, but these are imperfect measures. Integrating some economic performance data like the Am Law 200, for example, may at least enable us to answer normative questions about successful strategies.

Conclusions

In this Article we examine empirically a number of possible implications of the increase in corporate in-house legal departments and the consequent decline of relationship lawyering for the structure of the legal industry, the structure of law firms and the nature of legal careers. We find weak evidence of the demise of midsized firms at the expense of large firms and small boutiques, though significant evidence that the largest firms have grown substantially. We also document a significant increase in the number and geographic diversity of multi-office law firms. These trends are consistent with a move by firms to compete with in-house legal departments by offering a broader range of services, with greater depth, across multiple jurisdictions.

We also document how and where leverage-the number of associates per partner-has increased over the past seven years. Overall leverage has increased, though not among firms of all sizes or in all locations. Leverage is highest for the largest firms and smaller firms and slightly lower for midsized firms. It is also significantly higher for multi-office firms and firms with New York City offices. Leverage has been growing over the past seven years in small firms, but not in the large firms.

While the overall increase in leverage is consistent with the need for firms to retain younger lawyers to do the legal work that cannot be done by more senior lawyers, the patterns of leverage changes are not entirely consistent with this story. We would have expected to see this increase in leverage be at least as great in large firms as in small ones if the reason was the need to do the work that the rainmakers were bringing in.

Finally, we document changes in the nature of legal careers. We show that the average time to promotion-the amount of time between graduating from law school and being promoted-has increased by about half a year in the past seven years. We also present some weak evidence for a decline in the use of up-or-out promotion systems by law firms.

There has been a modest increase in the likelihood that an associate, if not promoted by the average time to promotion in her firm, will still be at the firm two years later. In addition, there has been some increase in the use of "off-track" lawyers in our sample of firms. Once again, these trends are broadly consistent with an increasing divergence between the job of a partner (rainmaking) and that of an associate (legal work) and an attempt on the part of firms to retain productive associates without promoting them to partner.

Despite some weaknesses in these data (the short time frame, our inability to distinguish between income partners and equity partners, the lack of data on economic performance), we believe that we have provided new data and some new insights on the possible effects of changes in the nature of firm-client relations. Our findings are broadly consistent with conventional wisdom about the effects of the decline of relationship lawyering, though there are some pieces of the puzzle that do not fit the overall picture. Of course, this may be because we are not looking over a long enough time period, or our data are insufficiently detailed. But we believe that we have at least opened up some new avenues for exploration.

Pre-Law at Duke

Being pre-law at Duke, "means being involved in a process of self-discovery, defining life goals, determining life style and developing a strategy for realizing these goals." Dean Gerald Wilson, Pre-Law Advisor.

The Pre-Law Advising Centre sponsors programmes and distributes information to all students considering careers in law. Underclassmen who have specific questions that are not covered by Pre-Law Handbook or other resources may meet with the Pre-Law Advisor during walk-in hours posted early in the semester. You can also learn more about law school and the law through the student organization, Bench and Bar, which offers a variety of programmes and activities for those considering law as a profession.

Merging Across the Atlantic

For UK firms looking to expand across the Atlantic, a merger maybe the smartest move. Freshfields Bruckhaus Deringer announced at the start of 2006 that it is hoping to merge with an American law firm in the next five years. New York is the only significant financial market where the firm has yet to establish a sizeable presence and the firm says that the only way to do so is to combine with a US law firm.

Freshfields, along with Clifford Chance, Linklaters and Allen & Overy, is one of a so-called Global Quartet of UK law firms. Clifford Chance merged in 2000 with New York firm Rogers & Wells—a merger that was beset by difficulties and is only now settling down after a series of departures. Allen & Overy and Linklaters have both decided to build their New York offices by hiring lawyers from US firms, despite the fact that they can rarely offer them more money than they are earning elsewhere—a hard sell at best.

To date, the transatlantic mergers of law firms have usually involved at least one second-tier law firm, with the deal by DLA being the first true merger of equals, albeit outside the Magic Circle and Wall Street domain. Chicago firm Mayer Brown & Platt took over City corporate boutique Rowe & Maw in 2002 and in 2003 Cleveland's Jones Day took over another small City firm, Gouldens.

While there is still no shortage of American law firms looking to break into the European market by acquiring an existing, established practice, most firms prefer to take over smaller ones and impose their own corporate culture. The top US law firms are both much more profitable and much smaller, than the top English firms. Skadden Arps is the largest, with 1,600 lawyers, compared to 3,300 lawyers at the UK's largest firm Clifford Chance. The average Skadden partner earns $1.7m, while at Clifford Chance the average is $1.2m.

If Freshfields succeeds in convincing a top-tier New York firm that it wants to be taken over by a larger, less profitable, European suitor, it will undoubtedly lead to a raft of similar deals, as clients reconsider their options.

Consolidation, Transformation

For now, the most likely law firm mergers are likely to be domestic or pan-European. These are easier to achieve and there are many firms outside the Magic Circle who feel they simply need more resources to compete with their larger rivals.

If that isn't enough to make the London legal market a darn interesting place to be for the next few years, consider that the government is planning some pretty serious reforms of the rules governing law firms. Parliament asked Sir David Clementi, a former chairman of the Prudential, to review the market for legal services last year and he made some serious recommendations which look set to totally transform the way law is practiced in the UK.

Chief amongst Clementi's suggestions: a change to the current law which allows only lawyers to be partners and owners of law firms. That means that big businesses, including retailers like Tesco, might buy a law firm so that they could offer their customers legal services, in addition to the host of insurance and financial services products they currently sell. It also means accounting firms might team up with legal firms to offer a one-stop shop advisory service to clients., or law firms could make their finance directors partners. The permutations are limitless. Even more fundamental, is the prospect that law firms might one day float on the Stock Exchange and thus raise cash from external shareholders to make investments. The government is still deciding precisely what form the new rules will take, but there's no doubt that big changes are on the way.

Competition from US Law Firms

While the "Magic Circle" of large London law firms may be returning to form after a rough patch during the economic slowdown of the first years of the century, they now view their competition in Europe as the US firms that have built significant presences here—firms like Latham & Watkins, White & Case, Shearman & Sterling and Skadden.

Historically, a second tier of major UK firms would have challenged the Magic Circle for big deals, but those practices—Norton Rose, Lovells, Simmons & Simmons, Denton Wilde Sapte and Ashurst—are finding it ever harder to compete.

It is that second tier of UK law firms that has been hardest hit by the arrival of American law firms. They have lost some of their best-performing partners and, in many cases, have been unable to compete like their Magic Circle colleagues. This group of firms is also squeezed from below by the national UK law firms, which can compete by offering more competitive billing rates to clients and are frequently more entrepreneurial and aggressive in addition to being tightly managed.

The likes of DLA—which merged in 2005 to form DLA Piper Rudnick Gray Cary, the third largest law firm in the world—Eversheds, Addleshaw Goddard and Pinsent Masons, now give the London firms a run for their money.

National goes International

DLA is an interesting case in point. The firm began in Sheffield and as recently as five years ago was regarded as a regional law firm with ambitions to make its mark on London. That is precisely what DLA achieved—the firm's business model is based on hiring a few high quality partners, paying them a lot of money and getting the work done by more junior people often operating out of less costly regional business centres.

Such leveraging, as it is called, may concentrate the power in the hands of the few, but it satisfies the clients, who get a cheaper and more commercial service. DLA's combination with two US law firms in 2005 takes the model one step further. While it may not compete with firms like Clifford Chance and

Skadden Arps for M&A work, it will pick up global mandates from clients looking to use one firm worldwide for all sectors.

But size isn't everything and the firms at the polar opposite from DLA have done well by staying small and focusing on filling a niche. This group in London includes Macfarlanes, S.J. Berwin and Travers Smith, three corporate firms that continue to attract some of the biggest domestic work around and Berwin Leighton Paisner, which is arguably the top law firm in the UK for real estate work and has now added a strong corporate and finance business on the back of that brand.

These firms do not have the costly international networks of their larger rivals, but instead work in collaboration with independent firms in other jurisdictions when they must. This strategy, dubbed a 'best friends' model and demonstrated on a larger scale by the strategy of Slaughter and May (the only Magic Circle law firm without an overseas network), is attractive to certain clients. Macfarlanes, for example, last year advised its long-standing French client Pernod Ricard on a £7.6bn takeover of UK rival Allied Domecq.

Lawyers are in Demand in the UK

The London legal market is buzzing again and young lawyers are once more in high demand. In 2000, a global salary war originated in Silicon Valley, when law firms hiked wages to prevent associates from fleeing to dot-coms offering big perks and even bigger bonuses. The fever spread to New York, where attorneys were leaving for the investment banks and then to London.

When the market went quiet, associates were worried about losing their jobs, rather than inadequate salaries. Now that firms are busy, they are once again short of staff. The battle to hire the best lawyers and the brightest graduates entering the profession, has arguably never been tougher, as this time around there are not just UK law firms but sizeable UK offices of American law firms looking to hire the best people.

Salaries are rising again, but wage is not the only factor driving job-seekers' decisions. Firms now look to attract people by showing that they take things like diversity initiatives, corporate social responsibility and flexible working policies, seriously—things that were neglected at the expense of profit drives over the last few years.

The pressure to keep profits up means the chances of making partnership in any law firm are now scarcer than ever. Law firms operate an up-or-out policy—you join a track to the partnership and each year you are appraised to see whether you are still in line for the prize. Fewer and fewer make it now, but for those who do, the lucre is filthy.

The American law firms in London offer a whole new set of options for young lawyers. They pay more, no doubt and whilst they still have a reputation for longer hours, it's now fair to say that all serious law firms in London bow down to the almighty billable hour.

Smaller offices—most US firms still have less than 100 lawyers in the UK—mean more responsibility earlier on, but they can also mean a shortage of resources and partner time for tutoring. The training schemes may be less developed, but the work is just as good. English law firms are another matter entirely and have a huge battle on their hands: to attract the best lawyers, to retain their market share in their home market and to win on the incredibly competitive global legal stage.

Overview of Major Practice Areas in the UK

Is M&A Back? : The second half of 2005 and early 2006 have seen the first signs of serious mergers and acquisition (M&A) activity reappearing in Europe since the downturn that began in 2001. When M&A and capital markets went quiet following the dot-com crash and the 11 September attacks, English law firms, like so many businesses around the world, found things difficult.

After a decade of unprecedented expansion both in their London offices and their international networks, firms suddenly found that they had a lot of lawyers with very little to do. That's an expensive overhead and through a combination of redundancies and aggressive performance reviews, the numbers of fee-earners were driven down to more manageable levels.

Law firms in London, wherever headquartered, are cautiously optimistic that the market is only going to improve. M&A seems to be coming back, with big deals like Spanish telecoms group Telefonica bidding a grand £17.7bn for English mobile company.

The merger, if and when executed, would break records as the largest takeover of a UK company since 2000. There are also rumours of a multi-billion quid bid from another Spanish company for UK airports operator BAA. Meanwhile, UK gas group BOC has been the subject of a $14.9bn bid from Germany's Linde. More recently, ports and ferries group P&O had been at the heart of a bid battle between rivals in Dubai and Singapore. Last year, however, DP World (the same Dubai Ports involved in the recent American uproar over ports control), agreed to a £3.92bn takeover of P&O.

Capital Markets Work

The other big corporate work for law firms is in capital markets, which has been a dismal state of affairs since 2000. With the world's two largest capital markets in New York and London, it's readily apparent why law firms are so keen to establish themselves in both markets. But a confluence of ill-fated events—the dot-com bust, the 11 September attacks and widespread corporate chicanery a la Enron caused many investors' cash and confidence levels to drop. Since Enron's collapse, the American Securities and Exchange Commission introduced tough compliance and reporting rules for public companies in an attempt to avoid similar blow-ups from happening again.

The measures have been a mixed blessing: increased corporate confidence but more red tape. The so-called Sarbanes-Oxley Act in particular makes it a lot more expensive to sell shares in the US and has thus benefited London by ushering businesses to a more forgiving market. As testament to that fact, the number of flotations in London has just returned to something approaching 2000 levels, with 421 IPOs (initial public offerings) in 2005, compared to 486 in 2000. There were only 97 in the year 2002, but with many Russian, Israeli and Eastern

European companies now choosing to list in London rather than New York, the UK exchange is again as busy as ever.

Also active on the corporate side is private equity work, with a huge number of private equity houses now investing in ever-larger European businesses. Private equity firms take money from institutional investors and then buy established companies with a view to improving their performance.

They then either sell them or take them public in hopes of delivering huge returns to the institutional investors. Since it began as a niche practice during the '80s, this business has developed enormously, with a number of large American private equity firms opening European operations and buying businesses here. Late 2005 saw the record-breaking $12bn bid for Danish telecoms group TDC, the largest acquisition by these buy-out groups ever seen on this side of the Atlantic.

Other Practice Areas

For all the sizeable English law firms, the return of corporate work is the most important development for their practices since 2000, as it is the mainstay of their revenue and profit generation. Other departments, like real estate, banking and litigation, are less cyclical and often less remunerative than the big M&A and capital markets work.

That said, business is now booming across the board for most firms. The property and projects teams are kept busy through a combination of building work for the Olympics (coming to London in 2012—sorry Paris) and the arrival of a new UK law that introduces property investment vehicles to England for the first time. These so-called Real Estate Investment Trusts, or REITS, are immensely popular in the US and continental Europe.

The litigation market is also busy, with the UK emerging as a major centre for international arbitrations and the increasing prospect of U.S.-style class actions coming to Europe. Cases like the Railtrack litigation, where a group of shareholders came together to sue the UK government over its handling of the company's demise, are seen as the forerunners of group actions this side of the pond.

The Increasingly Competitive London Legal Landscape

The City of London's legal market is arguably the most competitive in the world. Ten years ago, there were only a handful of top law firms populating the City. However, with London's ever-increasing significance as a global crossroads, things have changed quite a bit over the past decade.

The key moment came in the mid-90s, when British law was modified to allow American lawyers to go into partnership with English lawyers. Since then, American firms have been flooding in and show no signs of slowing down. While a few English law firms might harbour ambitions to open offices in New York, their colonial proclivities pale in comparison to the Americans', as the number of American firms doing business in London now tops 100. The top 50 of these, according to London law magazine Legal Businesss, now employ more than 3,000 lawyers in the capital alone—2,200 of them English qualified. It's no secret that the aggressive American firms have a comparative advantage: a lot of money to throw around.

The six so-called Magic Circle law firms in London are, in order of turnover, Clifford Chance, Linklaters, Freshfields Bruckhaus Deringer, Allen & Overy, Slaughter and May and Herbert Smith. These are the firms widely recognised as dominating the corporate work and advising on the highest value litigation and capital markets transactions. All six have had to focus on raising their profitability levels in the wake of the economic downturn following the dot-com crash and the terrorist attacks of 11 September, not least to keep their partners happy in the face of tempting offers from the Yanks. The result has been a paradigm shift: like their U.S. counterparts, there is an increasing focus on the dreaded billable hour. Lawyers are expected not only to log the hours, but to be on call for clients day and night. Annual targets of 2,000 billable hours are not unusual as the days of long lunches and carrying less productive colleagues have become a thing of the past.

The Americans can afford to pay more because they are, by and large, more profitable businesses. A part of the reason for American firms' relative success is that they haven't made the extensive overseas investment that many top English law firms have. A larger reason lies in the U.S. litigation culture and its mind-boggling revenues. Of course UK firms do litigation, but it's a smaller part of their business, constituting 10-20 percent within the Magic Circle. This would be highly unusual for a major American firm, as it is typical for half of a U.S. firm's lawyers to work solely on the contentious side of the business. Of added significance is that litigation is not only profitable, but counter-cyclical;, when capital markets and M&A were quiet at the start of this century, American law firms saw no dip in revenue while UK firms suffered. Even when businesses aren't merging, absorbing and raiding one another, companies love to sue—which means a steady flow of cash to the firms representing them.

Last year was a banner one for U.S. firms doing business in London. According to The Lawyer, virtually every American law firm saw solid to stellar performance in '05. Debevoise & Plimpton, for example, posted an unprecedented 67.5 percent increase in revenue coming out of the London office (accounting for 10 percent of Debevoise's total revenue in 2005). LeBoeuf Lamb Greene & MacRae reported similarly impressive earnings, marking 2005 with a 43 percent rise in UK revenue. Meanwhile, Shearman & Sterling's UK partners took home, on average, £950,000—not a great leap over 2004's figures, as The Lawyer reports, but still 20 percent higher than the global average for profits per equity partner (PEP). Dechert partners also had some cause for celebration, as 2005 saw the firm's UK revenue climb by 15 percent and UK PEP leap nearly 30 percent to £846,500. White-shoe Wall Street firm Simpson Thacher boasted a 30 percent rise in UK revenue. Impressive numbers like that (with equally impressive profits per equity partner) will surely demand the attention of both English lawyers and their employers—English firms.

When it comes to the war for talent in the London market, U.S. firms take a different approach to partner compensation, which has attracted some of the England's top advocates. As a rule, Americans pay partners according to how much work they generate (a.k.a. the 'Eat what you kill' rule). The largest UK firms, on the other hand, pay partners according to seniority.

This means that the highest-billing at English firms (rarely the most senior, rarer yet the highest-paid) would be better off parlaying their professional zealotry into higher earnings at a U.S. firm.

There are other factors luring English attorneys to American firms, of course. As The Lawyer reported in March 2006, the recent defection of two high-level Linklaters private equity partners to Chicago-based Kirkland & Ellis seemed linked to tension between Linklaters' private equity team and its banking group. Kirkland, operating through a headhunter (who, according to The Lawyer, blabbed about the move months in advance), capitalized on Linklaters' internal strife, taking Graham White and Raymond McKeeve—along with a portable book of clients including CVC and TDR Capital. The move demonstrates that American firms—which, being comparatively new to the market, are freer to try new things—may be more willing to accommodate disgruntled attorneys' professional whims than traditional UK firms.

All of this means that for a talented lawyer entering London's legal market, U.S. firms offer a viable—and rather lucrative—alternative as a place to begin a career. However, you would be gravely mistaken to think you can make a killing working at an American firm without putting in American hours. As a result of the whizbang billable ethos, all of the law firms in London (American and English alike) now expect their associates to earn their keep and then some.

Qualifications need to be a Legal Secretary: Those interested in starting a career as a Legal Secretary would need to have completed a secretarial course. A good legal secretarial course will train you to perform highly specialised work requiring detailed knowledge of legal terminology and various procedures pertaining to the practice of law. When you are starting out, the most important thing to have is a good typing speed.

In addition, a legal secretary should have the ability to pay close attention to detail; have an organised approach; the ability to work under pressure; the ability to work on their own initiative as well as being discrete in order to deal with confidential information.

Salaries: Salaries vary depending on a secretary's level of experience as well as the skills that they can bring to their future employer.

Benefits and Perks: Benefits vary from firm to firm but over all they include the following:

Health Insurance: Health insurance is considered to be one of the primary benefits on offer for candidates. Group schemes are available whereby employees pay monthly subscriptions through their wages and receive a 10% discount. A number of the larger firms would offer to fund or part-fund health insurance for their staff.

Pension Schemes: Pension schemes are relatively uncommon for support staff in the legal industry. Those firms, which do provide a pension scheme usually, reserve it as a benefit for employees who have been with the firm for a number of years (*i.e.* after two years of service).

Holiday Entitlement: There is huge variation amongst firms with regard to holiday entitlement. Most firms offer the standard 20 days per annum with some firms escalating to 23 days per annum. A common trend in the provision of benefits is compensation for length of service or loyalty to a firm. Many firms offer additional holidays to employees after 2/3 years with the firm.

Flexi-time: Flexi-time is an appealing but unusual benefit in the legal industry. It affords employees the opportunity to work extra hours, which can be paid as overtime or taken off as time in lieu. Only a handful of firms offer flexi-time.

Over-time: Most firms would pay overtime to support staff as the legal industry is a pressurised environment, which requires the flexibility to work outside regular working hours.

Travel Pass: This is rapidly becoming one of the most popular benefits available for support staff in the legal industry. Depending on the firm some employers offer to pay for the monthly travel pass whereas in other firms the travel pass goes through an employee's wages thus becoming a tax efficient method of payment.

Annual Review: Most firms would review salaries on an annual basis. A failure to perform reviews promotes discontent, which in turn encourages turnover.

Educational Reimbursement: Some of the larger firms offer educational reimbursement if the employee is pursuing a course, which is related to his/her position within the firm. Upon completion of the course the employee may be of additional benefit within the firm *i.e.* a legal secretary who completed the diploma in legal studies may be able to run some files under supervision.

Christmas Bonus: The Christmas bonus is a discretionary benefit, which varies from firm to firm and from employee to employee.

Attendance Bonus: The attendance bonus is a monetary bonus which an increasing number of firms are offering support staff for consistency in attendance levels.

Reduced Legal Fees: Depending on the firm some employees may be offered reduced legal fees for the sale or purchase of property. Smaller firms tend to offer this service to employees more so than larger firms.

Promotion Prospects: Legal firms tend to promote staff from within once they have proved their ability to manage files independently. For those working as a Legal Secretary who would like to move up the career ladder, many complete a Diploma in Legal Studies, which will aid them in gaining a position as a Legal Executive. Many Secretaries who receive solid training decide to further themselves within the Legal sector by qualifying as a Solicitor.

What the market is like for Legal Secretaries/ where the jobs are: Legal Secretaries are in a favourable position as there is a constant need for their skills in the Dublin market at present. There is always a consistent demand for experienced Legal Secretaries particularly right now in the area of Commercial Property and Commercial work. This is due to the high levels of construction in Ireland as well as the increase in companies setting up in Ireland. The Legal Division at Meghen Group has first hand experience of this growth as we have four consultants that recruit exclusively for Legal Support positions for permanent, contract and temporary roles. There appears to be renewed energy in the industry, which has encouraged some firms to expand. This expansion of course creates additional opportunities, which prompts movement on the candidate's side.

Lifting the Blinkers—Life Outside London

It may come as a surprise to a few of our younger readers, but there is legal life outside the capital. Whilst the very largest deals and biggest salaries remain the preserve of the Magic Circle and the City's increasing horde of US firms, there is also a very healthy living and excellent quality of work to be found outside EC1. Increasingly, high profile corporate transactions, the larger commercial deals and more complex property developments are being farmed out by cost conscious blue chip companies to the bigger regional and national law firms; recognising the latter's increasing sophistication and capacity to deal with this type of work.

Salaries Outside of London: Like the quality of work, salaries outside London are also rising. Firms recognise that in order to attract (and just as important retain) the top level lawyers required to carry out this deluge of instructions, they need to pay commensurately. As a result, last year saw newly qualified salaries in many regional practices break through the £30,000 barrier for the first time and it is not unheard of for equity partners in the likes of Wragges, Walker Morris, Eversheds, DLA and Pinsent Curtis to draw upwards of £400,000 a year. Admittedly not the headline grabbing sums linked to some London practices which elicit gasps of surprise amongst young lawyers and of concern amongst paying clients, but nonetheless, not a bad living.

The concentration of financial institutions and blue chip HQ's will always mean that the very largest deals and most eye-popping pay reviews will tend to gravitate to London, but this is sometimes at a cost. High salaries have to be financed by some means and often this is done by imposing high chargeable hours targets on fee earners; sometimes in excess of 1800 hours per annum. This compares with targets of 1200-1300 hours as is commonplace in many of the major regional practices. It's all very well earning huge sums of money but it's also nice to have spare time in which to spend those hard earned bank balances. Cynics argue that regional firms' insistence that they offer a better quality of life is simply a euphemism for the fact that they do not pay as well as their London competitors. Others would say there is some substance to the cliche.

Benefits of Working outside London: As one would expect, house prices are also lower outside the capital. Exorbitant property prices in London consign many junior solicitors and barristers to years of rented accommodation until they are able to save the substantial deposits required to make a first step onto the property ladder. In contrast, the regional pound buys a lot more bricks, mortar and surrounding land—again another lifestyle issue.

This piece is not intended as an exhaustive review of opportunities in the regions, nor is it going to portray a provincial solicitor's working life as a 9 to 5 bed of utopian roses. Work-related stress, long working hours and traffic congestion still blight the lives of lawyers in the regions as they do in London but are perhaps a less frequent occurrence than is the case with their City counterparts. The intention is simply to give a flavour for regional life and to give an overview of some of the principal players in the legal market in the main commercial centres outside London.

Conclusion: Life outside of London is not for everyone. Many people thrive on the hustle and bustle of life in the first city; the eye-catching nature of the transactional work, the sky high salaries, the proximity of so many top class restaurants and cultural opportunities. But these are not exclusive to London. True, they may exist in greater concentration than in Birmingham, Leeds, Bristol and the like but they're still there. In the final analysis, people most likely to look for a move from London to the regions are those who have a pre-existing tie to those particular localities be it through friends, family or historic university links.

What they should take comfort from is that the regions are able to offer the whole spectrum of legal opportunities from small high street general practices to well regarded, major commercial firms offering high quality work and competitive salaries. Whatever it is you are looking for, a suitable opportunity should exist, in one of the other commercial centres in England and Wales. You may not become an overnight millionaire but then again you may enjoy a better overall quality of life.

What is Competency based Interviewing?

Competency based recruitment is a style of interviewing.

- What is it?
- How can I prepare for it?
- How will I be measured?

What is It?: It's a style of interviewing used so that a candidate can best show how they would demonstrate certain behaviours/skills in the work place; by answering questions about how you have reacted to and dealt with previous work place situations.

By using past experience a potential employer can predict future behaviour by:

- Eliminating misunderstandings
- Preventing personal impressions
- Reducing the candidate's ability to "fake"

You will be asked to give an example of a situation or task which led you to take a certain course of action. Probing questions will then be used to determine the course of action you took and what changes were created by those actions and the effects of those actions on others.

Traditional job descriptions are now quite out-dated. Most organisations will analyse a role by breaking it down into key competencies. For example the competencies of a lawyer may be planning and organising, innovation, personal drive, problem analysis and decision making. If an organisation uses this type of interviewing, it is very likely that your job will be defined on this basis and your performance in it will be managed through competencies. For example your appraisal may well be linked to evidence and real-life examples of having demonstrated these competencies.

How can I Prepare for it?: The simple answer is you can't. You can prepare for the interview in the normal way by researching the company, making sure you are able to talk through the work and skills that you have described on your CV. You will be given the opportunity to market yourself for the role in the normal way.

However, most candidates will find this type of interviewing much more interactive and enjoyable and it is likely that the interviewer will be able to encourage you to be much more open than in a traditional interview.

How will I be Measured? : Usually the interviewer will have a number of pre-planned questions to ask you. For each of these questions they will ask for real-life evidence where you have demonstrated the behaviour or skills. They will know what the desired behaviours are and will look for positive and negative indicators. Examples of competencies and related questions are as follows:

What Salary can I Expect if Working in an Investment Bank as a Lawyer?: The last decade has been one of great change in the investment banking legal market. An increasing number of banks have come to realise that an in-house legal resource is not only a valuable strategic addition to a company, but an effective cost reduction centre, allowing for significant savings in external legal fees.

This realisation has been coupled with an increase in the number of lawyers seeking to move out of private practice and into investment banking, both in legal and non-legal roles (particularly corporate finance). This has translated into a vast increase in both the number and calibre of in-house lawyers, particularly in the last four years.

This maturing of the in-house legal market has also meant that increasingly banks are having to compete with various other sectors seeking to attract high-quality lawyers. Primarily these are UK and US law firms, with the impact the latter have had on UK salaries also having filtered down to affect the in-house salary market. Just as UK firms have had to increase salaries to try and combat the influence of the US firms, so banks have had to make their packages more attractive in order to lure away well-paid UK lawyers from private practice.

These trends are reflected in the figures contained in this year's banking legal salary figures. Base salaries have risen greatly in the last year and are generally comparable with those in top City law firms. In addition the in-house banking lawyer expects a healthy bonus in at least the 30-50% range (and often far higher).

Salary Figures

Banking and finance	*Min*	*Max*	*Average*
00Qualifier	£40,000	£50,000	£45,000
99Qualifier	£46,000	£52,000	£49,000
98Qualifier	£55,000	£67,000	£60,000
97Qualifier	£60,000	£76,000	£68,000
96Qualifier	£71,000	£80,000	£75,000
95Qualifier	£75,000	£87,000	£82,000
94Qualifier	£78,000	£92,000	£87,000
93Qualifier	£82,000	£95,000	£90,000

Why might a Company Recruit their First Lawyer?

What Value will I Add to the Business?

- By setting up a panel and managing the external law firms the company currently instructs. The trained legal professional in-house lawyer will be able to ensure the legal advice the company purchases is cost effective and by providing a great deal of the initial advice yourself, you will be able to reduce the company's spend on external lawyers.
- Working alongside the business people, an in-house lawyer can utilise the collective experience within the company to implement standard precedents, ensuring consistency and efficiency.
- You are likely to also resume responsibility for all regulatory issues. A good in-house lawyer will always

have one eye on potential future legal issues and problems and by planning ahead will be able to ensure such issues either do not arise or, if they do arise, that they are managed appropriately. External counsel, by definition, are one step removed from the business and cannot function in this way.

- By monitoring employment, pensions, litigious, property and company secretarial issues. These are typically areas of work which can be bought immediately in-house at a much lower cost.
- An in-house lawyer can add value in drafting and negotiating commercial agreements as well as on corporate transactions. As an employee, you will have intimate knowledge of the business. This familiarity will produce dividends in the time saved explaining the issues facing the business to external lawyers and results in contracts which fit the needs of the company much more closely.
- Having one point of contact for legal issues rationalises the company's approach to legal matters, saving time and money.

Typical Objections

Will there be enough work? Potential legal issues are often put to one side unless critical; it is only when an in-house lawyer is bought on board that it becomes apparent how much legal work there really is. In-house lawyers will be able to deal with employment, pensions, insurance, litigation, property and company secretarial matters as well as the core commercial issues arising daily in the business.

Why pay for an in-house lawyer when companies still need to employ external lawyers? The work may be able to be done entirely in-house. If not, an in-house lawyer will be able to select the appropriate law firm and the individual lawyers with the right level of experience to handle the issues as they arise. They will be able to monitor their work and fees charged and manage all internal advice to the company, therefore saving time and money. How does a company justify the up front expense on the annual budget? In the long-term an in-house lawyer will more than justify their salary by foreseeing the future problems, keeping the company up to date with regulatory issues and undertaking day-to-day legal work themselves.

In-house lawyers are increasingly self-sufficient, computer literate and able to work without support staff. Their expense would be budgeted throughout the year and it would prevent massive, unforeseen spending on solicitors fees paid for by the hour.

If a company hires an in-house lawyer, will I be able to continue to outsource the legal work that I don't want to do myself?

Most lawyers enjoy the legal work they undertake and whilst they may be interested in a semi-managerial role, will recognise a need to undertake day-to-day legal tasks themselves, rather than drawing on their budget for external law firms. This is particularly so in smaller companies where there is just one in-house lawyer.

Sample Job Description

Recruiting Administrator/Executive Assistant

FLSA Status: Non-exempt

Reports To: Office Administrator, Hiring Chair, Summer Associate Committee Chair

Summary: Provides administrative support in the areas of office programmes and events, corporate communications, attorney staffing programmes and the summer associate programme.

Essential Duties and Responsibilities include the following; other duties may be assigned as needed):

Coordinate Lateral Attorney Hiring:

- Work with the hiring chair and practice group chairs to coordinate recruiting efforts.
- Coordinate interview schedules, travel, evaluation reports and follow-up with candidates.
- Draft offer letters and compensation packages for associate candidates.

- Correspond and provide service to new hires, including assistance with relocation and moving, bar exam & review applications and materials.
- Develop and maintain working liaisons with recruiters.

Coordinate Summer Clerks and New Associates Hiring:

- Maintain contact with law school placement officials, including scheduling and providing necessary paperwork, plus hosting events for law students.
- Administer the interviewing process, including arranging travel and interview schedules, expense reimbursement and follow-up.
- Coordinate Hiring Committee, including setting meetings, drafting agendas, preparing materials and reports.
- Maintain database information on applicants.
- Project and oversee budgets.
- Prepare and update recruiting information for ads, websites, surveys, brochures, etc.
- Maintain contact with and provide service to new hires, including assistance with relocation and moving, bar exam & review applications and materials and bar stipends.

Summer associate programme. Work with Summer Associate Committee to plan and administer programme, including:

- Coordinating summer schedule.
- Planning and organizing orientation.
- Enlisting associates and partners to participate as mentors and writing coaches.
- Communications with office re: programme.
- Planning and organizing training.
- Preparing and distributing materials.
- Preparing and overseeing budget.
- Planning and overseeing events.
- Setting weekly meeting agendas.
- Overseeing work assignment and evaluation process.

- Working with committee to troubleshoot and counsel as necessary.

Support Associates Committee as Needed, Including:

- Coordination of new attorney orientations.
- Associate training.
- Exit procedures.

Provides Administrative and Secretarial Support to the Office Administrator, Including:

- Coordinating meetings.
- Making travel arrangements.
- Opening and routing mail.
- Prepare expense reports.
- Typing—preparing charts, agendas and minutes.
- Filing
- Providing a variety of administrative information to staff and attorneys in all offices.

Coordinates Office Events, Including:

- Assist in coordination of firm-wide attorney retreat.
- Coordinate staff appreciation week and office-wide lunch.
- Coordinate long-term employee event.
- Coordinate annual partners dinner.
- Coordinate miscellaneous office functions, as needed.
- Assist Benevolence Committee.

Qualifications:

- Ability to read and write English, including the ability to compose memoranda, agenda, manuals, etc.
- Ability to understand and follow work directions.
- Keyboard skills and familiarity with computers and business software, including Microsoft Word, Outlook, Access, PowerPoint and Excel.
- Ability to pay close attention to detail and be highly organized and accurate.
- Ability to communicate verbally in a professional and courteous manner.

- Willing and able to work in a team-oriented service environment.
- Public relations skills and ability to represent firm in a professional manner.
- High school diploma or GED Certification.
- Minimum of three to five years office experience, preferably in a law office.

Additional Requirements:

- Interaction and communication with others.
- Shift work and/or overtime.
- Fast work pace, multi-tasking and/or frequent deadlines.
- Highly repetitive work and/or following established procedures.
- Attention to detail, close concentration and accuracy.
- Professional office attire, appropriate to the position.

Recruiting Director

Reports To: Executive Director/Hiring Partner

Status: Non-Exempt

Classification: The position of Recruiting Director is a full-time position within the firm. The individual assigned to this position will receive benefits afforded full-time employees as set forth in the Staff Policies Manual.

Hours: 8:30 a.m. - 5:00 p.m. Monday through Friday Strict adherence to these hours is required.

Essential Functions: Interact with law schools to facilitate fall interviewing process; prepare necessary reservation forms; provide firm descriptive information.

Make all travel arrangements for on-campus and in-house interviews.

Design and direct in-house interview programme.

Ensure that all appropriate correspondence is produced and that appropriate records are maintained.

Write and place advertisements for attorney positions.

Handle employment inquiries for all lateral attorneys, new

and summer openings and arrange preliminary and full-scale interviews, as appropriate.

Prepare, administer and audit recruiting budget.

Design and administer summer associate programme, including organizing and attending social events.

Update NALP form and related firm materials (work with Marketing Department on recruiting-related promotional materials, including recruiting portion of firm's website).

Serve as firm contact with all outside organizations involved in the legal recruiting process, including the National Association for Law Placement, other law firms and legal search consultants.

Understand and be familiar with current industry and market trends and philosophies.

Administer attorney Continuing Legal Education as follows:

Input and manage database tracking information;

Run quarterly status reports for firm's attorneys;

Coordinate payment of course enrollment fees and any associated costs;

Enroll attorneys in programmes; and

Maintain hard copy files for auditing purposes.

Track all associates' vacation accrual and run quarterly reports.

Assist with selection, administration and evaluation of new attorney orientation materials. Plan and coordinate comprehensive semi-annual [or as needed] new attorney orientation.

Assist Executive Director as follows:

Assist in event planning;

Assist in project-oriented tasks such as surveys, etc.

Additional Duties:

- Attend training, education and development programmes to assist in improved performance, personal growth and advancement.
- Any and all other duties as may be assigned by supervisor(s).

Experience/Knowledge Required: Should have high school education or equivalent. Three or more years' experience in a legal environment preferred. Requires excellent organizational skills, attention to detail and flexibility with diverse personalities. Professional appearance and demeanour and ability to maintain confidentiality essential. Ability to use Microsoft Office 2000 and a document management system (DMS) preferred, as well as knowledge of database management. Good knowledge of English language, grammar and spelling. Ability to accurately proofread documents. Ability to set priorities and work under pressure.

Equipment Used: Personal computer/printer; typewriter; copy machine; facsimile machine; telephone; dictaphone; adding machine.

Director of Lawyer Recruiting

Department: Organizational Development

Reports to: Chief Organizational Development

Receives Work Direction From: Hiring Chair, Management Committee, Policy Committee

Supervises: Recruiting Coordinator, Recruiting Assistant

Provides Work Direction: Recruiting Coordinator, Recruiting Manager, Office Administrators

Duties and Responsibilities: Manage firmwide Recruiting Department. Provide day to day work direction and professional development, conduct periodic performance evaluations and annual reviews for the department. Responsible for hiring, training and evaluation members of the department.

Formulate, administer and maintain recruiting policies, operational procedures and practices consistent with the goals and objectives of the firm.

Manage recruiting processes and implement support and information sharing systems. Guide and manage efforts to implement new technology, software programmes and more efficient processes.

Work with group heads and office heads to develop and implement appropriate recruiting strategies and searches.

Work with group heads and office heads to negotiate compensation and other offer terms for candidates.

Develop and manage annual recruiting, diversity and attorney functions budgets. Investigate, analyze and budget for new recruiting initiatives; develop and analyze expense reports.

Oversee summer associate and fall recruitment programmes in all offices. Work with Managing Partner annually to determine summer associate capacity and fall associate capacity.

Develop, prepare and analyze reports and projections regarding attrition, market analysis hours and billing rates.

Serve as main contact for the firm for law schools, search firms and industry organizations (NALP, TCPRA, etc.).

Identify legal requirements and regulations affecting recruiting policies and procedures and ensure compliance.

Responsible for attorney offer process, including offer letter language, uniformity, due diligence, attorney registration and liability issues.

Oversee arrival and exit procedures for all attorneys. Conduct exit interviews and summary reports.

Work with Diversity Partner to implement firmwide diversity initiatives.

Serve as a member of the firmwide service initiative team.

Oversee non-partner attorney requests regarding transfer/ office relocation.

Negotiate, review and approve contracts, retainers and exclusivity agreements wit search firm agencies and outside vendors.

Prepare reports regarding the impact of firm's long- and short-term goals for growth retention and diversity.

Establish and maintain strong working relationships with other departments including IS, Finance, Human Resources, Professional Development and Marketing.

Conduct new attorney orientation programmes and oversee and organize firmwide mentor programme for first year associates and develop integration tools for lateral hires.

Develop appropriate interview techniques and evaluation forms for the various categories of attorneys.

Develop and maintain working relationships with law schools' career placement offices.

Develop and manage collateral recruiting materials, including development and updating of print materials and content of recruiting section of firm's web site.

Knowledge and Abilities: Required:

- BA degree and 5+ years experience in recruiting, career development or marketing.
- Ability to organize and implement detailed and complex projects.
- Excellent interpersonal skills and ability to deal effectively with all levels of personnel.
- Ability to work independently under pressure.
- Ability to handle a wide variety of duties.
- Excellent oral and written communication skills.
- Experience with PC and spreadsheet applications.

Director of Legal Recruiting

Reports to: Executive Director

Basic Function: Working with the Executive Committee Chair and the Managing Partner, draft and establish a strategy for lateral hiring. Under the direction of the Executive Director and the Chair of the Hiring Committee and coordinating closely with Practice Group Leaders, determines the need and coordinates the recruitment of associate attorneys and law clerks. Has primary responsibility for orientation, training and mentoring of attorneys. Oversees the summer clerk programme. Provides limited assistance to the Associate Review Committee and the HR Director in the annual associate review process and maintenance of files regarding performance reviews.

Responsibilities and Duties

Essential Functions:

- Recruiting
    - o Communicates regularly and specifically with Practice

Group Leaders regarding lawyer and paralegal needs and recruiting efforts to address those needs for both short and long-term.

- o Delegates most long-term (new associates) needs recruitment to Legal Recruiting Coordinator to pursue with law schools and clerking programme.
- o Coordinates short- and mid-term associate recruitment needs and approach with Chair of Hiring Committee once needs assessment is thoroughly vetted with Practice Group Leaders.
- o Uses available direct recruitment tools—phone calls, ads, introductory meetings, internal queries, etc.—to identify and attract proper candidates for each assessed practice need.
- o Arranges appropriate interviews with strong candidates with Practice Group Leaders and Hiring Committee Chair.
- o Manages the above lateral recruitment process to conclusion. Delegates all day to day administrative duties to the Legal Recruiting Coordinator. Supervises the efforts of the Legal Recruiting Coordinator.

- Training
  - o Develop and implement a firmwide attorney training programme, to incorporate general firm knowledge elements, combined with practice-specific training elements.
  - o Work closely with Executive Director and other senior directors and the Firm's General Counsel to develop and implement the general firm knowledge training sessions.
  - o Coordinate efforts with the Practice Group Leaders to develop and implement the practice-specific training sessions for attorneys in their respective practice groups.
- Mentoring
  - o Coordinate with the Hiring Committee Chair to develop and implement the key elements of a firmwide associate mentoring programme.

  - o Formalize the institutional aspects of the mentoring programme, coordinating with the HR director as appropriate.

Ancillary Functions:

- Work closely with other senior directors in coordinating the recruiting function with the other operational areas of the firm.
- Meet with the Chair of the Associate Review Committee to facilitate the initial transition of those duties to the HR Director.
- Maintain active membership, with firm support, in the appropriate legal associations, such as NALP, DALRA, ALA, etc. to ensure currency of "industry" norms, benchmarks, practices, etc.

Qualifications:

- Four-year college degree preferred.
- Knowledge and skills gained through experience and demonstrated ability to use knowledge effectively in the execution of tasks.
- Prior legal recruitment experience in a law firm preferred.
- Experience with human resources information systems and applicant flow tracking.
- Strong communication skills.
- Ability to establish effective working relationships throughout the Firm.
- Ability to work effectively under time constraints.
- Willingness to meet the requirements of the position, accepting a wide variety of assignments as called upon.

Director of Professional Recruitment & Development

Classification: Officials and Managers

Status: Exempt

Reports to: Managing Partner, Chairpersons of the Hiring Committee, Associate Review Committee and Professional Development Committee, Director of Administration.

Responsibilities: Manage all aspects of legal recruitment, development and evaluation of all non-partner attorneys. Firm representative to law schools, law students, search firms, NALP and other professional organizations.

Summer Programme:

- Direct all aspects of the summer associate programme.
- Maintain regular communication with all summer associates prior to arrival.
- Administer summer associate orientation.
- Plan and coordinate summer associate training and mentoring programmes.
- Maintain daily contact with summer associates. Oversee work assignments and work allocation, work with assignment coordinators of various practice groups to ensure sufficient availability of open assignments through the summer.
- Plan and organize all social events.
- Oversee the mid- and end of summer evaluations and exit interviews.
- Assist in assessing practice group hiring needs and matching summer associates with individual practice groups.
- Prepare, maintain and monitor summer programme budget.

Recruiting Programme: Schedule the fall on-campus calendar. Review schools visited on an on-going basis and recommend changes as appropriate.

- Pre-screen all unsolicited student resumes and respond to all applicants.
- Work closely with the hiring committee to schedule on-campus interviews and coordinate all callback interviews. Prepare data for hiring committee and schedule meetings for fall recruiting decisions.
- Facilitate decision making process and follow-up.
- Prepare and analyze reports and projections regarding law school recruiting success and other issues for use by committee.

- Conduct interviewer training on an annual basis.
- Update NALP forms and complete other necessary forms and surveys for law schools.

First Year Programme:

- Oversee all aspects of new attorney arrival, orientation and initial development.
- Administer first year mentoring and training programmes.

Lateral Hiring:

- Oversee all aspects of the lateral hiring process. Screen all unsolicited resumes and respond to all applicants. Solicit practice group interest, extend invitations to interview and maintain regular status report.
- Work with practice group leaders and managing partner to identify staffing needs.
- Communicate with executive search firms regarding lateral hiring.
- Facilitate decision making process and follow-up, including offer letters and necessary reference checks.
- Oversee all aspects of lateral associate orientation.

Associate Evaluation and Compensation:

- Direct the administration of the associate evaluation process. Ensure full communication with committee members and attorneys regarding evaluation policies and procedures.
- Serve as a member of the Associate Review Committee and make recommendations on compensation and progression issues.
- Develop, prepare and analyze reports and projections regarding compensation, hours, profitability and other issues for use by committee.
- Assess industry data regarding associate salaries.
- Oversee post-evaluation meetings with associates.
- Evaluate effectiveness of evaluation process and recommend changes as appropriate.

- Maintain and update associate evaluation software, including creating reports, troubleshooting and working with programmers on new features for future releases.

Professional Development Programme:

- Direct the administration of associate professional development in the areas of mentoring, work allocation, review and training. Ensure full communication with committee members and attorneys regarding professional development programmes.
- Serve as a member of the Professional Development Committee and make recommendations on training, mentoring and associate review processes.
- Work with practice group leaders to design and implement training programmes on an on-going basis.
- Develop and oversee budget for continuing legal education.
- Conduct informal exit interviews with departing associates.

General:

- Develop, maintain and oversee attorney recruitment materials.
- Participate in professional association activities for recruitment and development and participate on professional panels and committees.
- Develop and maintain relationships with law schools' career services offices.
- Prepare annual budget for recruitment and development of associates, including bar memberships and travel expenses.
- Recommend changes to the Professional Recruitment pages on the firm's intranet.
- Provide general counseling to associates to resolve job-related concerns.

Qualifications:

- College degree.
- Demonstrated leadership skills such as organizing, coordinating, problem solving, decision-making and interpersonal skills necessary for effective management.

- Good computer skills, including the ability to maintain the associate evaluation software. Minimum 3-5 years legal related experience.

Professional Recruitment Manager

Reports To: Office Hiring Partners, Human Resources Director, Office Managing Partner

Scheduled Hours: 9:00 am to 5:30 pm; or as otherwise negotiated with supervisors

Status: Exempt

Year-Round Responsibilities:

- Supervise all legal recruiting activities for area offices:
  - o Fall on-campus recruiting
  - o Summer associate programme
  - o Lateral hiring
  - o Lateral shareholder hiring (report to Office Managing Partner)
- Manage recruiting staff, including training, performance feedback & counseling.
- Membership in Hiring, Summer Associate & R2D2 committees; support committees.
- Oversee compilation and/or preparation of demographic information and hiring statistics.
- Prepare Recruitment Department Annual Budget which includes transportation, moving and bar expenses for all new attorneys, on-campus recruiting and summer programme.
- Keep abreast of current recruiting practices and maintain membership with the local city group and the National Association for Law Placement (NALP).
- Membership in Firmwide Recruiting, Retention, Diversity and Development Committee.
- Manages transition of new associates entry into firm.
- Manage offer and follow-up process.
- Manage marketing of firm to recruits.

- Manage FW OCI budget.
- Manage FW recruitment donation budget.

Winter/Spring Responsibilities:

- Oversee Summer Planning: includes verbal and written contact on a regular basis with summer associates regarding arrival/departure dates, housing needs, assignment preferences, travel arrangements.
- Prepare legal training programmes for summer associates with Director of Professional Development.
- Together with Coordinator, prepare materials for Summer Associate Committee meetings.

Summer Responsibilities

- Provide maximum support to summer associates to ensure that their experience at the firm is of the highest quality.
- Act as troubleshooter for any issue that summer associates may encounter.
- Attend all work assignment subcommittee meetings.
- Monitor summer associate work flow; track evaluations of work products.
- Participate in mid- and end of summer reviews for summer associates.
- Supervise and assist in preparation of offer decision meeting held by Hiring Committee and Department Chairs.
- Approve all disbursements for travel and social/ miscellaneous summer expenses.

Fall Responsibilities:

- Supervise preparation of all materials for on-campus interviewers.
- Supervise all callback invitations for schools visited.
- Together with coordinator, screen all unsolicited resumes for qualified candidates attending schools we do not visit.
- Together with Coordinator, organize and attend all Hiring Committee meetings/discussions/offer decisions for next summer programme.

- Act as liaison between the firm and candidates by personally meeting them to ensure personal contact with future summer associates.
- Attend weekly offer decision meetings.
- Interview candidates and provide feedback to Hiring Committee.

Lateral Recruiting:

- Establish and maintain relationships with legal search firms. Work with legal search firms and advertise available positions in legal newspapers.
- Screen all resumes, in coordinator's absence.
- Recommend potential candidates to practice group leaders and hiring partners.
- Together with Coordinator, act as liaison to lateral candidates throughout the interviewing process.
- Determine new and cost effective methods regarding lateral searches in a competitive legal market.
- Work with Professional Recruitment staff in other offices to develop and promote Firmwide recruiting efforts.

Retention Issues:

- Work with R2D2 to determine projects. Assist in implementation of same.
- Prepare orientation & training programmes with Director of Professional Development and HR staff.

Firmwide Issues:

- Oversee Recruiting Intranet site.
- Work with other offices to create standardized procedures.
- Work with Director of Human Resources to standardize policies in all offices.
- Oversee FW summer associate summit.
- Oversee FW OCI process.
- Prepare FW OCI training.

Requirements:

- 5 years experience in professional recruiting.
- Bachelor's degree or equivalent or experience.

Legal Recruiting Manager (LRM)

General Summary: Overall management of firm-wide lateral recruiting, summer, associate development and diversity programmes. Overall administration of department. Member of the Recruiting Committee.

Immediate Supervisor

Committee Chairs

Chief Administrative Officer

Immediate Subordinates

Special Projects Assistant

Recruiting Committee:

- Oversee development, implementation and updates of hiring policies and procedures with the Firm's Recruiting Committee.
- Coordinate Recruiting Committee meetings and determine agenda.
- Produce Monthly, Year-End and ad hoc Recruiting Reports.
- Oversee production of recruitment materials.
- Direct information management including record keeping, statistics and reports.

On-Campus Interviews:

- Coordinate entire on-campus interview process firm-wide:
  - o Work with CRC and RC to select appropriate law schools.
  - o Screen resumes; work with attorneys to select interview candidates.
  - o Coordinate attorney on-campus interview assignments.
  - o Work with attorneys in all offices to determine office interviews, summer offers and rejections.

Summer Associate Programme:

- Coordinate annual summer associate programmes for all offices:

- o Manage calendar of summer associate social events firm-wide and coordinate summer associate social activities in this office.
- o Coordinate summer associate training programmes to educate summer associates on the practice of law, skill training and recruiting.
- o Work directly with CAO, CRC and Human Resource Manager on personnel issues and crisis control regarding summer associates.
- o Maintain the Clerk Review Evaluation Worksheets (CREWS) database and control flow of assignments to summer associates in this office.

- Work directly with CRC to analyze regional and national summer associate salary scales to determine a competitive salary structure for all summer associates.

New Associate Preparation:

- Maintain practice group assignments for PGLs.
- Coordinate pre-arrival activities including bar review, bar exam and related issues, term sheets and advances.
- Arrange Orientation Training and serve as point of focus for new attorneys.

Lateral Attorney Recruiting:

- Coordinate year-round lateral hiring for all offices. Work directly with Firm managers (MP, PGLs and PICs) and RC Members in each office to analyze hiring needs.
- Work directly with CAO to negotiate headhunter contracts and act as point-person for all lateral placements and outplacements through headhunters; place advertisements and coordinate all lateral recruiting trips.
- Coordinate technology programmes to maintain firm-wide attorney recruitment database and Internet to enhance recruiting.
- Screen all unsolicited resumes and respond to all applicants. Solicit departmental interest and extend invitations to interview.
- Schedule and participate in candidate interviews, ensure attorney participation.

- Prepare term sheets and provide due diligence.
- Support diversity hiring goals.

Budget Monitoring:

- Track all recruiting budgets summer associate programme, on-campus recruiting, summer associate salaries, summer associate entertainment, bar review/ exam expenses, associate training programmes, etc.
- Review all associate training and recruiting-related expenses, code with appropriate accounting file number, enter into appropriate expense database and submit to CRC or Associate Liaison for approval.
- Review budgets on a monthly basis to determine compliance with budget.

Attorney Training and Development:

- Work directly with CRC, CAO and Associate Liaison to study retention process and methods to attract and keep quality candidates through internal public relations programmes and team building functions.
- Work directly with CRC, CAO, Associate Liaison and CFO to analyze regional and national legal salary scales to determine competitive salary structure for new associates.
- Work directly with Associate Liaison to coordinate in-house training programmes and educational programmes on a firm-wide basis, utilizing internal and external resources; monitor and obtain feedback from associate participation and progress in training and assist in assessing and improving associate training programme.
- Arrange Associate Meetings.
- Coordinate mentoring and new associate evaluation processes by working directly with Associate Liaison, PGLs and mentors.
- Coordinate and/or conduct exit interviews for all departing associates firm-wide and prepare periodic reports thereof for firm management.

Diversity Task Force:

- Work with Diversity Task Force Chair.

Policy Administration:

- Assist with the development, implementation and management of policies.

Relationship Building:

- Work directly with the Career Placement Departments at local law schools to conduct presentations with students on the recruiting, interviewing and hiring process.
- Maintain regular correspondence with Career Service Departments at law schools and the Legal Recruiting Departments at law firms. Organize annual meetings to discuss recent issues in the legal market. Organize receptions and events.
- Maintain membership in the National Association for Law Placement to remain current on the legal industry and recruitment.
- Coordinate with CMO and Marketing group for promotion of Recruiting Committee's efforts and improvement of communication.

Manager of Student Recruiting

Summary: Reports to the Director of Legal Personnel. Coordinates and manages all Firm student and new hire recruiting. Coordinates recruiting efforts with the Firm's Hiring Committee and regional Recruiting Committee members.

General responsibilities include recruiting, hiring, training and evaluation of summer associates and new hires. Assists in organizing and administering the summer clerk programme in each office. Travels to each region as needed to address recruiting and retention concerns.

Essential Duties and Responsibilities: "Essential Duties and Responsibilities" are primarily job functions that the employee must be able to perform unassisted or with some reasonable accommodation and with minimal supervision.

- Arranges for on-campus interviewing dates, schedules Recruiting Committee members for interviewing, prescreens resumes for the Hiring Liaison, schedules callback interviews and handles all communication and correspondence necessary in the process including offer letters.

- Organizes and administers the summer clerk programme in each office. Coordinates work assignments and social functions in all offices insuring the programmes are of the highest caliber, appropriate for the various markets, consistent in all locations and successful in attracting the best students who then accept the Firm's full-time employment offer.
- Serves as the primary contact for summer associates and new hires and coordinates orientation and training with office managers and key Firm personnel.
- Prepares recruiting budget for summer associates and new hires and approves all appropriate cash reimbursement requests.
- Maintains strong relationships with law school placement offices and other law firm recruiting managers.
- Completes NALP forms, law school surveys and any other statistical information required.
- Inputs all appropriate information for students and new hires into the Firm's recruiting database software.
- Researches and communicates new and successful recruiting techniques and other issues of interest to students and new hires.
- Participates in Hiring Committee and Recruiting Committee meetings as requested.
- Various other assignments relative to recruiting as requested.

FLSA Status: Non-Exempt

Working Conditions: Normal law office environment with little exposure to excessive noise, dust or temperature.

Human Resources/Legal Recruiting Administrator

***Reports To*:** Office Administrator, Hiring Partner

Department: Human Resources

Job Purpose: Provides administrative support in Human Resources as it relates to non-legal staff recruitment, compensation and counseling as well as legal and non-legal orientation, benefits and overall employment practices. Responsible for all legal recruitment functions including on-campus and in-house interviewing of law school candidates and lateral hires.

Screens and evaluates write-in candidates' resumes. Liaison between legal search firms, law school placement officers and Recruiting Committee. Provides Recruiting Committee current information with respect to status of job market. Coordinates and oversees summer associate programmes including researching and arranging for travel and housing; coordinates office assignments and secretarial support; schedules social events; compiles evaluations.

Responsibilities and Accountabilities:

1. Coordinates staffing needs; recruits, interviews and determines offers for non-exempt personnel; assists with recruiting of exempt personnel.
2. Develops and conducts orientation for staff, attorneys and summer associates.
3. Conducts staff counseling and terminations. Conducts exit interviews for staff and attorneys.
4. Indirectly supervises non-legal personnel in conjunction with Office Administrator.
5. Negotiates fee agreements with and procures services of employment agencies.
6. Administers recognition programme for staff; orders and arranges delivery of various gifts and acknowledgments.
7. Administers and conducts 90-day reviews for staff.
8. Administers annual non-exempt staff evaluation and compensation programme. Analyzes salary surveys, recommends compensation structure and determines increases and reclassifications for approval by office and firm administration.
9. Maintains files and assists with annual audit of the firm's Affirmative Action Programmes.

10. Participates in the development of firmwide policies and procedures.
11. Answers questions regarding firm policies and benefits including all insurance, health and dependent care spending accounts, 401(k) elections, partner physicals, real estate/will reimbursements, workers compensation claims and other related benefit matters. Administers flexible benefits programme and conducts meetings concerning programme changes.
12. Approves paid time off (PTO) leave from office for non-exempt staff. Administers staff and attorney leaves as they relate to disability or Family and Medical Leave Act (FMLA) requests and provides interpretation as necessary.
13. Prepares annual budget for staff compensation and other human resources operating accounts for approval by office and firm administration. Prepares quarterly reports noting and explaining budget variances.
14. Assumes secondary responsibility for administration in Office Administrator's absence.
15. Supervises, directs and assigns projects to the HR Assistant as required.
16. Maintains all files and records regarding attorney hiring.
17. Investigates various forms of associate training and development to be considered in the office.
18. Member of the National Association for Law Placement (NALP) and represents firm at regional and national meetings.
19. Participates in firmwide meetings of Hiring Partners, Recruiting Coordinators, HR Administrators and local Recruiting Committee Meetings.
20. Prepares (opens, sorts and reviews) resumes received for entry into database. Prepares non-interviewed resumes for HR Clerk entry in database. Responsible for mail merges and preparation of response letters to interviewed applicants.
21. Assists in completing and submitting registration forms (including NALP forms). Helps coordinate on-campus interviewing dates with law schools.

22. Maintains communication with professional placement services engaged to identify qualified candidates to meet the hiring needs of the office for both lateral lawyers and for support staff.
23. Assists Associates with BAR/BRI course preparation expense reimbursements.
24. Assists in selection of Mentors for new incoming Associates.
25. Coordinates annual Associate evaluation programme.
26. Processes Associate moving/relocation expenses.
27. Processes and distributes Associate sign-on and Judicial Clerkship bonuses.
28. Performs special projects as assigned.

On-Campus Interviewing:

1. Assists Hiring Partner and Firm Recruiting Administrator in determination of which schools to visit by evaluating the firm's past recruiting successes and/or failures at each school.
2. Assists Hiring Partner and Firm Recruiting Administrator in determination of potential attorney interviewers for each school and solicits attorneys' cooperation.
3. Acts as liaison between firm and law schools, for coordinating dates, logistics and scheduling of on-campus interviews, maintains communication with law school placement officers and prepares all relevant correspondence.
4. Completes and submits registration forms (including NALP forms) and firm literature required by law schools including relevant correspondence.
5. Coordinates travel itineraries and accommodations for attorney interviewers traveling to law schools.
6. Discusses candidates interviewed on-campus with interviewing attorneys and helps determines which candidates will be invited back to the office for further interviews.
7. Determines eligibility of unsolicited candidates' resumes and prepares responses.

Lateral Hiring:

1. Assists Hiring Partner and Firm Recruiting Administrator in determining lateral hiring needs. Consults with attorneys and provides feedback regarding team hiring needs.
2. After slot approval by Team Head, submits request for hiring approval to the Firm's Hiring Chairman on behalf of team making request.
3. Ensures that approval to use Hiring Agency is obtained.
4. Ensures that Hiring Agencies execute firm's fee agreements to maintain compliance under firm's Affirmative Action Plan.
5. Serves as office's liaison to headhunters engaged to meet hiring needs. Maintains communication with headhunters; obtains information regarding status of job market; provides and obtains feedback regarding candidates.

In-House Interview Process:

1. Schedules appointments for candidates interviewing in the office.
2. Arranges attorney participation in interview process and prepares and distributes interview schedule.
3. Receives each candidate upon his/her arrival to the office. Explains interview schedule to candidate and leads candidate to first appointment.
4. Meets with candidates to recap interviews and answer questions following full round of interviews.
5. Collects and compiles evaluations of candidates from lawyers for the Recruiting Committee.
6. Participates in decision-making process by providing feedback on candidates and evaluating feedback from interviewers.
7. Ensures that candidates are promptly notified of hiring decisions and prepares appropriate correspondence.
8. Assists Hiring Partner and Firm Recruiting Administrator in assigning and monitoring attorneys to "rush" candidates receiving job offers.

9. Arranges and monitors travel reimbursement of candidates interviewing in the office including coordination of shared expenses with other law firms.
10. Prepares statistical report regarding status of candidates.

Summer Programme:

1. Researches and coordinates Summer Associate and new associate housing options.
2. Prepares Summer Associate handbook, including information regarding firm policies & procedures, firm literature and any other information pertinent to the Summer Associates' employment with the firm.
3. Coordinates Summer Associate offices with Office Administrator.
4. Assists in the selection of Mentors for Summer Associates.
5. Assists Hiring Partner and Recruiting Committee in determining the schedule of summer events. Reviews former Summer Associates' evaluations of previous summer functions, budgetary restrictions, lawyer participation and timing restrictions.
6. Coordinates all summer events, including scheduling, reservations, transportation and attorney participation.
7. Encourages ongoing feedback from Summer Associates regarding the summer programme.
8. Compiles written evaluations of Summer Associates for use in formal evaluations.
9. Maintains communication with former Summer Associates to obtain information regarding their choice of employment.

Minimum Qualifications Required:

1. *Education:* College degree with emphasis on human resources administration (or related field) preferred or equivalent work experience.
2. *Experience:* Minimum of five years human resources (preferably in a legal environment) or strong administrative background including supervisory responsibilities. Previous work experience should include

employee counseling, employee performance assessment and compensation, policy administration, recruiting and employee training. Accounting background and knowledge of budget preparation helpful. Must be proficient on personal computer.

3. *Other Minimum Qualifications:* Responsibilities require strong written and oral communication skills, discretion, good judgment, tact and poise. To perform this job successfully, the incumbent must possess the abilities or aptitudes to proficiently perform each of the responsibilities and accountabilities described above. Position requires a high degree of tact and discretion. Major challenges of this position include the diverse nature of the responsibilities, effective communication of information to key personnel in a timely, accurate and sensitive fashion and adhering to multiple project demands at one time. Examples include: ability to work well under pressure, sound decision making capabilities, patience, diplomacy, high stress tolerance, excellent interpersonal communication skills, ability to prioritize workload and adapt to changing conditions. Visual acuity needed for transcription, extensive reading and input (typing) of information into computer. Must possess ability to receive and convey detailed information through oral communication. Essential competency processes include language ability, reasoning and memory.

Legal Hiring Coordinator

Status: Non-Exempt

Department: Legal Hiring

Reports to: Hiring Committee, Office Administrator

Supervises: Legal Hiring Administrative Assistant

Primary Responsibilities:

- Establish, implement and update policies and procedures for attorney recruiting and hiring, including new associates, lateral associates, staff attorneys, summer associates and law clerks for the office.

- Receive, pre-screen and process all inquiries regarding hiring of new associates, lateral associates, staff attorneys, summer associates and law clerks.
- Oversee preparation of correspondence in response to employment inquiries.
- Assist Hiring Committee in selecting law schools at which the office will conduct on-campus interviews. Handle registration for on-campus recruiting at selected schools. Assist Hiring Committee in setting hiring criteria for law schools at which the office recruits.
- Assist Hiring Committee in selecting attorneys who will interview on campus at each law school.
- Coordinate on-campus interviewing including preparing packets of information for interviewers to take to campus, briefing and debriefing on-campus interviewers, preparing follow-up correspondence to law students, etc.
- Coordinate callback interviews, including monitoring correspondence, ensuring that travel and hotel arrangements are made for the students and scheduling interviews.
- Negotiate hotel and relocation service agreements for recruiting and hiring purposes.
- Prepare weekly hiring status reports and maintain files on pending candidates for Hiring Committee meetings during the recruiting season. Monitor number of offers extended versus number of desired candidates.
- Ensure offer and rejection decisions are made and letters mailed on a timely basis. Conduct interviews of candidates who decline offers from the Firm to determine reasons and alternatives selected.
- Ensure the Firm's adherence to the guidelines of the National Association for Law Placement (NALP).
- Keep Hiring Committee informed about the status of recruiting efforts in the other offices of the Firm. Coordinate recruiting efforts with other offices whenever possible.

- Process recruiting travel reimbursement requests from students and other firms. Also process our requests to other firms for their share of a candidate's travel expenses.
- Annually revise and distribute surveys to law students who were extended offers during the recruiting season and to summer associates regarding the Firm's recruiting programme and Summer Associate Programme, respectively.
- Update and develop marketing materials for attorney recruiting (including Firm resumes, recruiting brochure, NALP form, etc.). Periodically review and revise recruiting form letters.
- Schedule attorneys to conduct in-office interviews with candidates for new associate, lateral associate, staff attorney, summer associate and law clerk positions.
- Serve as Firm liaison to law school placement directors and other recruitment professionals. Maintain these relationships through telephone contact and participation in professional organizations and conferences.
- Coordinate all aspects of lateral associate hiring, including working with practice group to devise search strategies, drafting and placing advertisements, working with legal search firms, pre-screening resumes and coordinating interviews, offers and rejections.
- Act as liaison between the Firm and legal search firms for the purpose of employing lateral associate attorneys.
- Verify newly hired attorneys' Bar admission and standing. Check references of law students and ensure that references of attorneys are checked by a partner.
- Ensure that all new hires complete a conflict of interest questionnaire prior to joining the Firm and have questionnaires reviewed by the Conflicts Department and Conflict and Ethics Committee.
- Keep Hiring Committee abreast of what other law firms are doing in terms of attorney recruiting, hiring and training.
- Plan and coordinate the Summer Associate Programme

including correspondence with incoming summer associates, organizing training seminars and social events and handling office logistics.

- Collect work assignments from attorneys and distribute to the summer associates. Monitor the summer associates' workloads through collection of Weekly Reports.
- Develop and implement orientation programmes for new associates, lateral associates, staff attorneys, summer associates and law clerks.
- Work with the Administrative Coordinator to the Executive Committee to plan and organize the Annual Firm-wide New Associate Orientation.
- Prepare legal hiring budget and other expenditure reports as requested.
- Prepare compensation surveys and statistical reports regarding attorney recruiting and the Summer Associate Programme.
- Coordinate exit procedures for departing attorneys and summer associates including conducting exit interviews.
- Revise, maintain and distribute the Associate Policy Manual.
- Assist the various practice groups' Training Committees with the planning and coordination of in-house attorney training programmes and collect information on and register attorneys for outside training and continuing legal education seminars.
- Assist the Lawyer Development Committee with any projects or tasks assigned from time to time.
- Assist the Office Administrator and Administrative Coordinator to the Executive Committee with any projects or tasks assigned from time to time.
- Assist the Paralegal Recruiter with paralegal hiring from time to time.

Recruiting Coordinator

- Primary responsibility for coordinating follow-up with candidates.

- Assigns and tracks follow-up with all candidates with offers.
- Coordinates all offer dinners and other recruiting special events.
- Schedules callback interviews.
- Greeter for on-campus interviews.
- Assists with law school presentations and programmes.
- Coordinates summer events—calendar and budget.
- Coordinates follow-up with summer associates.
- Facilitates summer events.
- Primary responsibility for development of recruiting materials, brochures and website.
- Assists with website enhancement for recruiting purposes.
- Collaborates with web editors to update internal and external website for recruiting purposes.
- Develops and updates materials for marketing to specific groups of recruits.
- Coordinates participation in surveys.
- Tracks and manages full-time hiring reports and data.
- Assists with surveys.
- Attends and participates in conferences.

Summer Associate Programme:

- Help coordinate April orientation for summer associates:
  - o Assist students with travel arrangements (air/ground).
  - o Prepare and distribute letters regarding orientation day.
  - o Coordinate lunch outside of office.
  - o Coordinate afternoon job fair with junior associates.
  - o Coordinate cocktail reception at the end of the day.
- Assist with preparation of rotation schedule to other offices.
- Coordinate various training programmes for summer associates.
- Prepare and maintain rotation schedule throughout summer programme.

- Prepare bulletin notices regarding arrival dates of summer associates.
- Arrange banking for the summer associates.
- Assist summer associates with obtaining a social security card (foreign summer associates).
- Prepare and maintain mid-summer/exit review schedule.
- Maintain review file for 50% of summer associates and monitor assignments on Lotus Notes.
- Prepare mid-summer/exit review memo and sit-in with hiring partner to deliver; prepare memo re: message delivered after review.
- Plan various social functions and local tours for summer associates.

Recruiting:

- Attend Recruiting meetings.
- Completion and submission of registration forms for Fall OCI/Job Fair.
- Completion of hospitality rooms requests/order refreshments.
- Host hospitality suite at various law schools.
- Schedule students for office interviews.
- Coordinate office interview (compile and distribute schedules to attorneys).
- Interview candidates and complete evaluation form.
- Prepare law school feedback forms (interview results from fall).
- Coordinate with various law firms for shared expenses by students.
- Arrange for various donations at schools (auctions, fund raising, student-sponsored events).
- Prepare various flyers for law school publications/events.
- Help coordinate firm-sponsored events.

NALP:

- Compile reports in order to update firm's NALP form online.

- Enter and submit pertinent information to NALP Organization.

Hiring Coordinator

General Responsibilities:

- Meet with hiring partner on a routine basis regarding hiring matters.
- Act as liaison between offices regarding hiring matters.
- Answer phone calls regarding employment inquiries.
- Respond timely to employment inquiries with a written response and maintain resumes on file (for one year).
- Maintain monthly calendar of all hiring activities.
- When recruiting lateral candidates, work closely with headhunters and act as liaison between them and the hiring committee.
- When necessary, coordinate the advertising of lateral positions in newspapers and law school publications.
- Assist new lawyers (entry-level and laterals) with relocation and associate entitlement matters (*i.e.*, salary advance, bar exam and moving expenses and general questions about benefits and procedures).
- Coordinate orientation schedule for new lawyers.

On-Campus Recruiting:

- In January/February of each year, assist the hiring partner with updates to the recruiting materials as follows:
  - o One-page firm resume;
  - o NALP Form; and
  - o Long firm resume.
- Maintain files on each law school with whom we recruit.
- Assist hiring committee in the selection of lawyers to send to on-campus interviews.
- Fill out on-campus interview registration forms for law schools, request checks for registration fees (March/April) and coordinate with on-campus interviewers regarding the selection of interviewees (September/October).
- Send out recruiting packets to all law review students prior to on-campus interview dates.

- Process response letters to law students after on-campus interviews and schedule callbacks.
- Make travel arrangements and hotel reservations for students scheduled for callbacks.
- Send out letters to law students confirming interview dates.
- Coordinate interview schedules (*i.e.* distribute interviewer availability sheets to all lawyers, make up interview schedule for each candidate and set up lunches and dinners).
- Greet candidates at the beginning of their interviews and give them a schedule of the lawyers with whom they will interview.
- Send out offer and rejection letters to callbacks.
- Produce weekly summary regarding status of callbacks for hiring committee meetings.
- Send statements out at the end of recruiting season to law firms in the city who invited back the same law students to invoice them for their share of the students' travel expenses.
- Produce final report for hiring committee members on hiring results and expenditures during in-campus recruiting.

Summer Associate Programme:

- Meet with hiring partner and summer associate coordinator to plan summer programme:
  - o Select advisors for each summer associate and write a memo to each advisor explaining his/her responsibilities;
  - o Communicate with summer associates regarding starting dates, salaries, overview of summer programme; and
  - o Assist summer associate coordinator in writing a memo to all lawyers informing them of starting dates of summer associates, requesting work assignments, soliciting participation in social activities, etc.

- Set up and maintain a personnel file for each summer associate.
- Assemble summer associate notebooks (include background on firm, practice group organization, personnel forms, timekeeping instructions, etc.).
- Coordinate with accounting office regarding billing rates and salaries for summer associates.
- Meet with firm administrator to make secretarial assignments and locate offices to accommodate summer associates.
- Submit newsletter announcements to firm's other office regarding introduction and starting dates of this office's summer associates.
- Coordinate with summer associate coordinator on getting ample work assignments from lawyers during summer programme.
- Maintain confidential notebooks on each summer associate which include completed work assignments and attorney evaluations on each work assignment.
- Coordinate with administrative assistant regarding "first day" matters (parking cards, keys, etc.).
- Coordinate orientation schedule for summer associates.
- Coordinate social functions for the summer programme:
  - o Set up a summer club membership for each summer associate;
  - o Plan outings to professional baseball games, concerts, theatre performances, etc.; and
  - o Plan welcome/farewell receptions and other office socials, etc.
  - o At the conclusion of the summer programme, produce a report on all summer expenses and submit to hiring committee.

Coordination and Planning Firm Holiday Functions

- Children's Holiday Party:
  - o Coordinate with Santa on a date for children's party; and

- o Send invitations, line up shoppers to buy gifts, decorate and order catering for approximately 50 children and their parents.

Recruiting Coordinator

Coordinate all aspects of the Summer Programme:

- Prepare for the arrival of Summer Associates including section assignments, arrival and departure dates and supervising attorneys.
- Plan social events, section parties, Summer Associate lunches and firm parties.
- Oversee evaluations for Summer Associates.
- Coordinate Summer Associate training and orientation.
- Conduct Summer Associate exit interviews.
- Monitor Summer Associate work projects and assignments.
- Arrange follow-up dinners and correspondence between Summer Associates and attorneys.
- Arrange in-home and other dinners at attorneys' homes.
- Prepare offer and decline letters for Associate and Summer Associate candidates.

Coordinate all aspects of the Fall Recruiting Programme:

- Plan pre- and post- on-campus interview (OCI) receptions.
- Arrange travel and interview schedules for candidates.
- Conduct correspondence and follow-up with candidates.
- Organize travel and schedules of interviewing attorneys.
- Plan fall OCI calendar.
- Coordinate with various law schools to schedule interview dates.
- Conduct orientation for new hires.
- Prepare National Association for Law Placement forms for all offices.

Assist in various Associate training and continuing legal education programmes.

Administer annual Associate reviews.

Assist in coordinating Associate retreat.

Coordinate firm-wide practice group section retreat.

Organize summer and fall recruiting receptions and events.

Arrange callback interviews for Summer Associate candidates.

Coordinate New Associate Orientation and training programmes.

Prepare annual recruiting budget.

Recruiting Coordinator

Summary: The Recruiting Coordinator assists shareholders on the Recruiting Committee with making arrangements for all areas of the recruiting programme for summer clerks and new associates.

Reporting Relationship: Reports and is responsible to the Shareholders on the Recruiting Committee, the Director of Human Resources and may report to the Managing Shareholder.

Essential Functions: Responsible for recruitment and orientation of summer clerks and new associates.

Responsible for setting up interviews at law schools.

Responsible for arranging in-firm interviews.

Responsible for screening and follow-up of write-in candidates.

Responsible for planning and running the Firm's summer programme for clerks, which includes, but is not limited to party planning.

Responsible for coordinating travel arrangements, hotel arrangements and in-office interview schedules.

Responsible for coordinating work assignments and the evaluation process for summer clerks.

Responsible for arranging the temporary housing for clerk's in the summer.

Responsible for handling the correspondence as it applies to the recruiting process.

Other duties as assigned.

Physical Requirements

Must have minimum physical mobility, ability to communicate clearly—both orally and in writing and to operate computer. Must have strong organizational, negotiation and communication skills. Must have planning and prioritizing skills. Must be able to work in stressful conditions under timelines. Must have strong interviewing and public relations skills. Must be able to travel. Must be physically present at work.

Recruitment Coordinator

Recruitment Committee:

- Liaison between multiple offices for Summer Associate Programme.
- Update Firm recruitment materials and NALP Questionnaire.
- Attend annual NALP Conference.
- Prepare OCI Registration Forms; schedule interview dates; contact person for law schools (20+ law schools, consortiums).
- Update and prepare materials for summer programme.
- Coordinate all areas of summer programme, including orientation, picture book, track work assignments, social events, evaluations, reviews, offers, update summer associate calendar, training programmes, attend events, etc.
- Prepare OCI materials for interviewers.
- Schedule callbacks; prepare interview materials; hotel/ travel reservations.
- Schedule and attend meetings, send notices, prepare minutes and callback reports.
- Prepare summaries of evaluations.
- Contact person by summer associates and new attorneys.
- Pre-screen write-in resumes.

- Select welcome gift items for summer associates.
- Coordinate 1L Holiday Reception; other receptions at law schools.
- Attend Diversity Committee meetings.
- Input resume data/activities.
- Prepare/provide any other requested material.

Professional Development Committee:

- Prepare annual training programme calendar.
- Schedule and attend meetings, send notices, prepare minutes.
- Schedule training programmes (date, notices, room, menu, etc.).
- Apply for CLE credit; provide notice for approval.
- Collect Form CLE 1 from attorneys; track timely filing.
- Contact person for CLE questions; provide CLE updates to attorneys (twice per year and as requested).
- Coordinate all areas of two-day New Attorney Orientation Programme.
- Develop programmes for new attorney continuing education sessions.
- Contact person for Mentor Programme (mentor/mentee assignments).
- Coordinate reviews of Partners.
- Orientation for new attorneys/paralegals.
- Prepare/provide any other requested material.

Associate Compensation and Evaluation Committee:

- Coordinate all areas of associate review process (for annual compensation adjustment and for mid-year review), including preparation of review calendar, update associate self-evaluation form and review form, distribute/collect associate self-evaluations, update associate files, prepare confidential binders for committee members, distribution of files, attend meetings, etc.
- Prepare/provide any other requested material.

Paralegal Review Committee:

- Coordinate all areas of paralegal review process.

Administration:

- Provide information to Executive Director and other administrative staff as needed.
- Track new hires/departures.
- Contact for Bar Exam questions.
- Contact for questions regarding moving companies/ expenses.
- Attend administrative staff meetings.
- Prepare biennial Affirmative Action Plan.
- Act as liaison between Hiring Committee, Professional Development Committee and Associate Compensation Committee.
- Prepare/provide any other requested material.

Recruitment Coordinator

Summary: The Recruitment Coordinator will assist with all areas of on-campus recruiting programme. This individual will also assist with management of the summer associate programme and have active involvement in lateral associate recruiting.

Essential Duties and Responsibilities include the following and other duties and responsibilities as assigned:

Receives, records and directs all inquiries regarding hiring of lawyers.

Organizes, for review and response, all materials, resumes, etc. received from headhunters and unsolicited applicants.

Assists with drafting/revising/updating correspondence, memoranda and other related documents as necessary.

Proofreads all documents and correspondence generated in the Recruitment Department to ensure accuracy and consistency.

Provides telephone coverage for Director. Screens calls, if requested and provides assistance. Assists with the handling of numerous calls by obtaining relevant information, returning calls, fielding questions, etc.

Maintains profiles and status of all recruitment activity in LawCruit.

Ensures that files are opened and applicants are entered into the LawCruit system.

Tracks and records all headhunter submissions and prepares resumes for review.

Corresponds with incoming applicants to schedule interview dates and times and inputs information into Recruitment Department calendar.

Schedules attorneys for in-office interviews with candidates.

When necessary, greets interview candidates in the office, informs applicant of process and provides brief description of each interviewer.

Conducts Bar verification for all newly hired attorneys.

Assists with filing.

Assists with the coordination and development of the summer associate programme, including social activities, training sessions, information dissemination, etc.

Assists with administrative organization of summer associate orientation.

Creates and prepares invitations to summer social activities, when necessary.

Maintains response lists and follow-up with guests and provides directions and coordinates rides.

Arranges OCI interview calendar.

Takes responsibility for the administration of the fall recruiting process. (Interfaces with candidates to set interview date and time, determines interests, determines what help candidate will need in travel and hotel accommodations, sets up interview schedule, follows-up with collection of reviews, compiles information and records for use in recruitment committee meetings, etc.)

Executes the distributing, collecting, compiling and assembling of all 1st year and lateral mid-year evaluations.

Assists summer associates in finding housing—send weekly mailing.

Coordinates and prepares Summer Associate notebooks for mid-summer and exit interviews.

Communicates and interacts in a friendly and professional demeanour with attorneys, staff and outside company representatives.

Ensures attention to detail.

Has good organizational skills.

Assists on special projects as assigned.

Adheres to Firm policies and procedures as outlined in Firm manual.

Acts as a role model within the Firm.

Maintains a positive and respectful attitude.

Communicates regularly with supervisor about department issues.

Dresses professionally.

Has a good attendance record.

Works overtime as required.

Is punctual and prepared to perform duties upon arrival.

Performs work honestly, ethically and safely.

Maintains confidentially.

Performs other duties as assigned.

Qualifications: To perform this job successfully, an individual must be able to perform each essential duty satisfactorily. The requirements listed are representative of the knowledge, skill and/or ability required. Reasonable accommodations may be made to enable individuals with disabilities to perform essential functions.

Individual must have strong organizational skills and ability to prioritize workload. Individual must also have the ability to handle: pressure; extended hours, if requested; interruptions; multiple projects with frequently changing priorities.

Education and/or Experience: Bachelor's Degree from four-year college or university and 1+ years of recruiting experience in a law firm environment, or equivalent education and work experience preferred.

Language Skills: Ability to read, analyze and interpret the most complex documents. Ability to communicate with candidates, attorneys, supervisors, colleagues and other staff members in a courteous and professional manner. Ability to respond effectively to the most sensitive inquiries or complaints.

Mathematical Skills: Ability to calculate figures and amounts such as interest, discounts, proportions, percentages, area and volume. Ability to apply concepts of basic algebra and geometry.

Reasoning Ability: Ability to define problems, collect data, establish facts and draw valid conclusions. Ability to interpret an extensive variety of technical instructions and to deal with abstract and concrete variables.

Equipment Used: Computer with Windows 2000 operating system (or higher), utilizing Microsoft Office software, including but not limited to Word 2000 (or higher) word processing software, Docs Open document indexing system, LawCruit, Outlook e-mail software and Excel spreadsheet software. Must also be able to use Lotus Notes, Lotus Approach and Lotus 123 software. Also must be able to master computer assisted research utilizing the Internet or other required software. Additional equipment used includes: typewriter, photocopy machine, facsimile, telephone communication system, postage meter and Equitrac cost recovery system.

Physical Demands: The physical demands described here are representative of those that must be met by an employee to successfully perform the essential functions of this job. Reasonable accommodations may be made to enable individuals with disabilities to perform the essential functions.

While performing the duties of this job, the employee is regularly required to talk and hear. The employee frequently is required to stand; walk; sit; use hands to finger, handle or feel; and reach with hands and arms. The employee is occasionally required to stoop, kneel, crouch or crawl. The employee must occasionally lift and/or move up to 35 pounds.

Specific vision abilities required by this job include close vision, distance vision, colour vision, peripheral vision, depth perception and ability to adjust focus.

Work Environment: The work environment characteristics described here are representative of those employee encounters while performing the essential functions of this job. Reasonable accommodations may be made to enable individuals with disabilities to perform the essential functions.

The noise level in the work environment is usually moderate.

Legal Career Coordinator

General Summary: Provides administrative assistance to the Recruiting Committee, Associate Liaison and Diversity Task Force.

Immediate Supervisor: Legal Recruiting Manager (LRM); Chairman of Recruiting Committee; Human Resources

Recruiting Committee:

Overall:

- Assist LRM with preparation of materials for Recruiting Committee meetings and Recruiting Committee development.
- Assist in creation and maintenance of statistical reports summarizing and tracking recruiting activities.
- Prepare NALP form and other law school surveys.
- Assist with Recruiting Committee special projects and development of recruiting procedures.

On-Campus Interviews:

- Enter law student candidate information into recruiting database and update database as students progress through the interview process.
- Prepare Interviewer Binders for attorneys conducting interviews.
- Prepare resume packets for Call Back Receptions.
- Update school placement office information; prepare informational statistical forms and hiring feedback reports for law schools.

Summer Associate Programme:

- Prepare Summer Associate orientation manuals.
- Prepare summer associate and new associate correspondence.
- Assist with Summer Associate Programme administrative tasks and special projects.

Lateral Attorney Recruiting:

- Continually maintain recruiting database, entering all attorney candidates firm-wide into the recruiting database and update database as applicants progress through the interview process.
- Maintain department information including applicant and correspondence files.
- Prepare offer and declination letters for CRC or LRM signature.

Budget Monitoring:

- Assist with budget reporting and statistical data.

Web Site Maintenance—Recruiting and Diversity:

- Assist with on-going updates to the recruiting and diversity sections of the firm's web site.
- Assist with e-mails to second-year law students directing them to Extranet.

Associate Training and Development:

- Prepare manuals for Associate Orientation and Training.
- Make logistical arrangements for Associate training sessions and track session attendance.
- Assist with Associate Meeting planning.
- Track Associate Mentoring Programme Evaluation submission.
- Prepare rosters, charts and mailing lists.
- Maintain Associate Information Binder.
- Assist with Associate Development special projects.

Diversity Task Force:

- Assist with ad hoc projects and initiatives.

Attorney Information:

- Handle all attorney registrations and dues for memberships for the Notary Archives, ABA and LSBA.
- Maintain attorney biographical information including law school, advanced legal education and start date.

Professional Recruitment Coordinator

Reports To: Professional Recruitment Manager

Scheduled Hours: 9:00 am to 5:30 pm; or as otherwise negotiated with supervisor

Status: Exempt

Year-Round Responsibilities:

- Coordination of legal recruiting activities for area offices
    - o Fall on-campus recruiting
    - o Summer associate programme
    - o Lateral hiring
    - o Lateral shareholder recruiting (as directed)

Screen and evaluate resumes. Manage on-campus and in-office interviews. Arrange on-campus interviews; prescreen on-campus resumes where appropriate. Schedule and/or supervise scheduling of in-office interviews.

- Membership in Hiring & Summer Associate committees; support committees.
- Together with Manager, compile and/or prepare demographic information and hiring statistics.
- Keep abreas of current recruiting practices and maintain membership with local city group and the National Association for Law Placement (NALP).
- Support Firmwide Recruiting, Retention Diversity & Development initiatives.
- Quarterly updates to new hires.
- Oversee scheduling of in-office interviews for all attorney candidates.

Winter/Spring Responsibilites:

- *Coordinate Summer Planning:* includes verbal and written contact on a regular basis with summer associates regarding arrival/departure dates, housing needs, assignment preferences, travel arrangements.

- Assist in preparation of materials for Summer Committee meetings.
- Plan office space and coordinating secretarial coverage for summer associates.
- Assist in planning orientation and initial training schedule for summer associates.
- Discussions with IT Department regarding computer needs for new hires.
- Support social event coordinators, *i.e.* make initial plans for larger events, supervise payment of events, prepare invitation lists and monitor invitation process.
- Survey fall offerees re hiring process.

Summer Responsibilities:

- Provide maximum support to summer associates to ensure that their experience at the firm is of the highest quality.
- Oversee planning office space and coordinating secretarial coverage for summer associates.
- Plan orientation and initial training schedule for summer associates.
- Discussions with IT Department regarding computer needs for new hires.
- Support social event coordinators, *i.e.* make initial plans for larger events, supervise payment of events, prepare invitation lists and monitor invitation process.
- Assist inpreparation of offer decision meeting held by Hiring Committee and Department Chairs.
- Planning for fall arrivals: transportation, offices, buddy assignments, etc.
- Participate in mid and end of summer reviews.

Fall Responsibilities:

- Preparation of all materials for on-campus interviewers.
- Organize and attend all Hiring Committee meetings/ discussions/offer decisions for next summer programme.

- Attend weekly offer decision meetings.
- Survey Summer Associates re summer programme.
- Act as liaison between the firm and candidates by personally meeting them to ensure personal contact with future summer associates.
- Interview candidates and provide feedback to Committee.

Lateral Recruiting:

- Establish and maintain relationships with legal search firms. Work with legal search firms and advertise available positions in legal newspapers.
- Act as liaison to lateral candidate throughout the interviewing process.
- Work with Professional Recruitment staff in other offices to develop and promote Firmwide recruiting efforts.
- Prepare and maintain all tracking charts and hiring statistics regarding invited candidates.
- Schedule candidates for call-back interviews.
- Prepare offer letters.
- Insure proper transfer of recruiting files to Attorney Personnel Manager after acceptance of candidates.

Retention Issues:

- Assist manager in implementation of R2D2 programmes.
- Assist in preparation of orientation and training programmes with Director of Professional Development and HR staff.
- Oversee office's Results Based Counseling Programme.

Requirements:

- 2-4 years experience with campus or lateral recruiting, firm or school side.
- Bachelor's degree or equivalent.

Human Capital Manager

This position is located in the Federal Government Agency, Human Capital Office (HCO). At the agency, this employee is the Human Capital Office's liaison to one or more mission and/

or mission support units, providing advice and assistance in a wide range of human capital management areas.

Major Duties and Responsibilities

- Facilitates the delivery of human capital services to customer, including recruitment and placement, position classification and compensation, performance management, performance and learning, employee benefits, employee relations, employee recognition and employee assistance (crisis management, mediation, counseling, outplacement, etc.).
- Advises management on position management concepts such as workforce planning and job restructuring. Compiles timely and accurate information on skills possessed by staff and provides analyses for management's use in identifying critical skills needed, reducing skills gaps and facilitating targeted recruitment, training/ development, retention and outplacement efforts.
- Collaborates with other human capital staff to develop and implement communication strategies to build teamwork, support matrix management principles and reinforce the agency's shared vision and core values. Analyzes feedback from new, existing and exiting employees to ensure continuous human capital management improvement.
- Initiates special projects, conducts evaluation studies and recommends new policies and programmes for improving the competencies and performance capability of the workforce in assigned units.
- Coordinates with the centralized recruiting function to identify the best strategies for filling talent needs in assigned units. Follows up on placements with appropriate investments in education, training and other opportunities to develop and retain the best possible workforce.
- Provides expert guidance and advice on appropriate performance appraisal, expectation setting and performance feedback processes. Gathers and analyzes feedback on the meaningfulness and effectiveness of the performance management system.

- Develops new or modified work methods, organizational structures, staffing patterns and specific supplemental guidelines to provide uniform application in addressing complex human capital problems in assigned units.

Knowledge, Skills and Abilities

- Mastery of human capital management, such that the employee is routinely applying new and experimental theories and developing innovative solutions to problems.
- Work at this level requires the ability to evaluate complex legislation and project its impact on the agency's human capital initiatives.
- Requires the ability to plan, organize and direct efforts to persuade the agency's management to accept and implement recommendations that may have long-term implications for the human capital programme.
- Ability to communicate effectively orally and in writing by preparing materials and presenting briefings that are appropriate to the audience and convey the message in a logical and persuasive manner.
- Ability to define problems, identify and evaluate alternatives, assess impacts and arrive at conclusions when participating in projects or studies to improve programmes that involve long-range, highly complex and often sensitive subjects.
- Ability negotiate effectively with management to accept and implement recommendations, where the proposals may require extensive changes in established procedures, or may be in conflict with the desires of management.

Operational Independence

The supervisor only provides administrative direction, which is broadly defined and supported by broad policy statement and/or basic legislation requiring extensive interpretation and judgement on the part of the employee. The employee has full technical and programme responsibility in carrying out work. The employee is considered an authority therefore work is generally accepted without change.

Scope and Impact

The primary purpose of the work is to deal with new and innovative ideas related to human capital management and to resolve critical issues. The work has a major impact on the agency's ability to meet its goals, objectives and requirements for human capital. The agency's employee has contact with a wide range of the agency's staff including the Managing Director level. Employee also is routinely in contact with representatives from other agencies and the private sector.

Public Defender Service

Introduction: The Office of Human Resources is responsible for developing, organizing, administering and evaluating the agency's comprehensive human resources management programme designed to insure optimum efficiency and effectiveness in the use of human resources to achieve agency goals and objectives. The incumbent of the position serves as Legal Staffing Specialist responsible for developing, implementing, managing, coordinating and evaluating the agency's legal recruitment programme designed for recruiting, examining and selecting candidates for all attorney and law clerk positions throughout the agency.

Major Duties and Responsibilities: Develops, implements, manages and coordinates the agency's legal recruitment programme designed to identify, attract and retain a high quality and diverse legal work force that is capable of accomplishing the agency missions. Analyzes and evaluates the agency's current staffing situation including turnover, vacancies and current recruiting strategies to identify and forecast staffing problems. Serves as principal advisor to agency management officials responsible for providing advice and guidance on a wide variety of technical issues involving legal recruitment and placement and for troubleshooting and resolving difficult and complex recruitment and placement problems including defining issues, determining applicable precedents, laws, or regulations and identifying the most effective approach and techniques to validate the correctness and propriety of the actions taken.

Plans, manages and coordinates all aspects of the agency's attorney recruitment processes designed to identify and recruit candidates for staff attorney positions in the five legal services divisions of the agency. Develops and publicizes recruiting announcements in law and professional journals and selected online websites; screens applications for basic eligibility; and refers a roster of eligibles to the attorney hiring committee for interview. Serves as resource staff and principal staff advisor to the hiring committee on the candidate interview and rating/ ranking processes.

Develops, manages and coordinates the agency's law clerk recruitment programme designed to recruit a diverse and talented group of the top law students committed to a career as indigent defense advocates to work full-time and part-time throughout the year in the agency's law clerk programme. Advertises recruiting announcements in selected websites; actively participates in law school recruiting forums to identify candidates; conducts on-campus or online screening interviews to determine qualified candidates based on such factors as their demonstrated commitment to public interest, quality of writing samples, academic achievements, legal and courtroom experiences. Participates with and assists the Special Counsel of the Legal Services Division in conducting the processes of candidate interview and selection.

Represents the agency at national or regional meetings, conferences or forums dealing with legal recruitment and staffing programmes. Actively participates in such meetings and conferences, making presentations to explain and seek support for the agency's programmes and establishing and cultivating contacts for recruiting. Establishes and maintains effective working relationships with law school officials, representatives of bar associations, officials of legal services agencies and law student associations interested in the agency missions and programmes.

Advises all levels of agency management concerning employment policies and practices that affect equal employment opportunity and work force diversity. Provides advice and guidance to managers and supervisors in the development and implementation of affirmative action plans. Maintains effective working relationships with minority bar and law student associations to improve equal employment opportunity relationships and conduct targeted recruitment to attract more minority candidates for work force diversity.

Administers the summer intern programme; issues programme announcements; identifies and recommends selection of interns; participates in orientation for interns; arranges or participates in training of summer interns on selected topics, including Westlaw, Lexis, computer training, etc.; establishes and maintains records on the work assignment and evaluation of summer interns; and coordinates periodic status meetings, mid- and end-of-summer reviews.

Performs Other Duties as Assigned

Factors: Factor 1—Knowledge Required by the Position

Knowledge of and skill in applying, a wide range of HR staffing and recruitment principles, concepts, regulations, techniques and guidelines sufficient to plan, develop and provide comprehensive legal recruiting and staffing services to recruit for staff attorney and law clerk positions throughout the agency; plan and execute creative recruitment campaigns to locate and identify highly qualified candidates involving numerous public contacts and recruiting sources; and advise management on a wide variety of technical issues involving recruitment and placement.

Expert knowledge of various legal recruiting practices, law school career services, specialized legal recruiting databases and public interest employment programmes, as well as demonstrated experience working with law schools and various other forums to locate and recruit qualified candidates for legal positions.

Knowledge of and skill in applying, written and oral communication techniques sufficient to develop and deliver briefings, project papers, staff reports and correspondence to management to foster understanding and acceptance of findings and recommendations.

Factor 2—Supervisory Controls

The incumbent reports to the Director of Human Resources, who provides administrative and policy direction in terms of broadly defined programme objectives and priorities. As a technical expert in legal recruiting, the incumbent independently plans and carries out the assignment with a minimum of technical guidance. Completed work is normally accepted as being technically authoritative and is reviewed only for overall technical adequacy and potential impact on broad agency policy and programme objectives.

Factor 3—Guidelines

Guidelines include agency policy, regulations and procedures; legal recruiting handbooks; reference materials and manuals; precedents; etc. Available guidelines are often ambiguous and express conflicting or incompatible goals and objectives, requiring extensive interpretation. The incumbent exercises judgement and broad latitude to determine the intent of applicable guidelines; and develop or recommend policy or guidelines for the agency's legal recruitment programme.

Factor 4—Complexity

Work involves serving as legal staffing specialist for the agency responsible for developing, implementing, managing, coordinating and evaluating the agency's legal recruitment programme designed for recruiting, examining and selecting candidates for all attorney and law clerk positions throughout the agency. Specifically, the incumbent analyzes and evaluates the agency's current staffing situation to identify and forecast staffing problems; serves as principal advisor to agency management officials concerning a wide range of technical issues involving legal recruitment and placement; and plans and conducts external recruitment processes to identify and recruit candidates for staff attorney and law clerk positions throughout the agency.

Factor 5—Scope and Effect

Work involves developing, implementing, managing and coordinating the agency's legal recruitment programme designed to identify, attract and regain a high quality and diverse legal work force that is capable of accomplishing the agency missions. Work affects the overall effectiveness of all legal services divisions throughout the agency and has a direct effect upon agency mission accomplishment.

Factor 6—Personal Contacts

Personal contacts are with all levels of agency management; officials of law school career centres; officials of legal, professional and educational organizations at national and regional levels; legal recruiters of public defender agencies, legal services agencies and public interest organizations; job applicants and the general public.

Factor 7—Purpose of Contacts

The purpose of contacts is to plan, coordinate and conduct programme operations; provide advice and guidance in resolving problems and issues; and to influence and persuade management officials to accept findings and recommendations. Contacts usually involve active participation in conferences, meetings, or presentations involving issues of considerable importance to the agency.

Factor 8—Physical Demands

The work is primary sedentary, with special physical effort or ability required to perform the work.

Factor 9—Work Environment

The work is performed in an office setting.

Attorney Recruitment Assistant

Reports To: Professional Recruitment Manager; Professional Recruitment Coordinator

Scheduled Hours: 8:30 am to 5:00 pm or 8:45 am to 5:15 pm; overtime when necessary

Duties: Year-Round Responsibilities:

- Manages all recruiting files.
- Opens and distributes mail, prepares disbursements, logs expenses, maintains department subscriptions.

- Provides secretarial support, answers telephones, types cover letters and various forms.
- Closes out files for new associate hires, prepares and distributes hiring memoranda.
- Responsible for "processing" all incoming resumes which includes checking for prior contact, completing a resume cover sheet and passing them on to the Professional Recruitment Coordinator ("Coordinator") for review.
- Sends resumes to attorneys for review, as requested.
- Delivers all correspondence requiring attorney signature, picks up and mails same.
- Initial review of monthly general ledger, highlighting questions for manager, researches questions with financial services.
- Maintains Hiring Committee and Summer Associate Committee binders.
- Attends Hiring Committee Meetings; membership in Summer Associate Committee.
- Maintains FW recruiting site on firm intranet page.
- Types and distributes all interview schedules.
- Assists Professional Recruitment Coordinator ("Coordinator") in scheduling interviews and filling cancellations.
- Helps greet applicants upon arrival and takes them to interviews.
- Gathers interview evaluations and types summaries.
- Maintains various Committee Distribution lists.
- Additional projects as assigned by Manager & Coordinator and various partners and Committee members.
- Photocopying and departmental faxing.

Winter/Spring Responsibilities:

- Handles payment of recruit hotel accommodations and reimbursement of travel expenses when the Firm is sponsor firm.
- Types all on-campus interview forms and letters for

securing interview dates at law schools including OCI forms.

- Works closely with Conflicts Department to determine that conflicts forms have been received for all new attorneys and summer associates. Maintains conflicts information on summer associates.
- Prepares and updates summer associate telephone and address lists.
- Monitors the return of summer associate questionnaires and photographs for intranet site.
- Prepares and updates all summer associates reports which include arrivals and departures, telephone address list, conflicts check-list.
- Prepare packets sent to summer associate advisors and monitoring advisors to be sure timely welcome letters are sent to summer associates.
- Secures time keeper numbers, voicemail boxes, Equitrac numbers, signage, etc. for summer associates.
- Sends to travel agent a list of all summer associates who will be booking transportation to area offices.
- Prepares summer associate first-day orientation packets and secures conference room space for weekly orientation breakfasts.

Summer Responsibilities:

- Member of the Summer Associate Committee.
- Responsible for preparing memoranda distributed to all hands announcing summer associate arrivals and departures.
- Responsible for maintaining summer associate personnel files which includes submitting all Personnel Action Notices and checking immigration forms.
- Assists Coordinator in coordinating social events for summer associates.
- Maintains a log of all expenses billed to the summer associate department and coordinates with accounting periodic running of current summer associate budget.

- Assists work coordinators in distributing work assignment forms, delivering evaluation forms and other tasks as needed.
- Coordinate OCI signups for all offices. Coordinate provision of recruiting materials to law schools.

Fall Responsibilities:

- Prepares on-campus interview binders for attorneys interviewing at law schools. These binders include recruiting information pertaining to the firm, information pertaining to the specific law schools, resumes and evaluation forms, travel itineraries and on-campus contacts.
- Coordinates Federal Express/messenger deliveries of on-campus recruiting materials to law schools for interviewers.
- Assists in coordination and billing of on-campus receptions.
- Assists Manager & Coordinator in scheduling callback interviews.
- Assists Coordinator in filling schedules for in-house interviews of on-campus and lateral recruits.
- Helps greet recruits upon arrival and takes them to interviewers.
- Types and distributes interview schedules.
- Gathers interview evaluations (which includes calling individual attorneys and personally pick up evaluations, if necessary). Types summaries.
- Types agendas for hiring Committee meeting and assists Manager in preparing handouts.
- Manages recruits' interviewing files and travel expenses.
- Assists in the preparation of a variety of reports and statistics required by the Hiring Partners, Hiring Committee, or Firm Management.

Qualifications:

- Proficiency in computer systems.
- Type 65 WPM net.

- Knowledge of MS Word, Excel, Powerpoint a plus.
- Excellent grammar, punctuation & spelling skills.
- Good command of the English language.
- Experience in standard & filing systems.
- Recruiting experience or experience with LawCruit database a plus.
- Professional demeanour.

Recruiting Assistant and CLE Coordinator

- Coordinate in-office interview scheduling for law students and lateral attorneys.
- Assist in coordinating on-campus interviewing.
- Schedule and coordinate with attorneys and law school personnel for on-campus interviews at over 10 law schools nationwide.
- Assist in processing all recruiting related expenses.
- Organize over 40 Mandatory Continuing Legal Education programmes for attorneys.
- Coordinate new attorney training programmes and annual Mock Trial.
- Assist in planning and facilitating all aspects of Summer Associate programme.
- Plan and facilitate Associate Offsite.
- Assist in coordinating firm-wide parties and special events.
- Administer annual Associate reviews.
- Prepare annual Professional Development Budget.
- Send notice to attorneys for all Hiring Committee meetings; order lunches; attend all meetings; record minutes and distribute to all members.
- Prepare and coordinate split billing with other law firms in the city following fall recruiting.
- Prepare rejection letters for Summer Associate and lateral attorney candidates.
- Administer Summer Associate work evaluations and reviews.

- Maintain all recruiting files.
- Open all mail.

Lawycr Jobs

Lawyer jobs advertised in the JobServe Legal industry job listings are those UK legal jobs that regularly feature lawyer jobs, solicitor jobs, legal secretary jobs, conveyancer jobs and paralegal jobs across all disciplines for lawyer jobs. If you want to find legal work specialising in either, administration or personal injury then use JobServe for your job search for legal jobs in the UK and abroad. You will regularly find the latest legal executive jobs, commercial property lawyer jobs and probate lawyer jobs on the JobServe website job listings.

Lawyer jobs in the Legal industry job listings are wide-ranging and encompass many disciplines within the Legal profession including; Administration, Banking, Business, Civil Litigation, Commercial, Company, Construction, Conveyancing, Corporate, Criminal, Divorce, Employment, Environment, Family, Finance, Immigration, International, Litigation, Marketing, Matrimonial, Personal Injury, Planning, Private Client, Probate, Property, Sales, Tax, Technology. When you search for Lawyer jobs you can choose to select the Legal industry tick box to limit your job search to find UK Legal jobs. Then you can add the search filter "lawyer" to return adverts for lawyer jobs.

Within these Legal disciplines the following jobs in Law, are advertised by both corporate legal departments and law firms, are regularly featured in the job listings. Commercial Property Lawyer, Conveyancing Lawyer, Corporate Lawyer, Criminal Lawyer, Divorce Lawyer, Executive Lawyer, Family Lawyer, Legal Executive, Litigator, Litigation Lawyer, Partner, Personal Injury Lawyer, Private Client Probate, Probate Lawyer, Property Lawyer, Tax Lawyer.

2

Career in Nanotechnology

Nanotechnology Industries

Fields may include, but are not limited to the following: physics, chemistry, applied mathematics, computational biology, material science, computer simulation/design/modelling/science, electronics, engineering, biotechnology, genetics, cellular biology, protein engineering, encryption, drafting, biochemistry, biosciences, metallurgy, art and genetics. Often times, several labs at a university departments of the previously listed will combine to create nanolab's for research.

Studying Nanotechnology

Many students have asked what they should study to prepare for careers in nanotechnology. Giving a decent answer requires outlining the different fields of research that fall under the nanotechnology umbrella and describing the background knowledge required to work in them. It also seems wise to say something about the different levels of knowledge and modes of learning that are relevant to such a broad, interdisciplinary area. The following is a personal view, based on what I have learned (and wished I had learned) and on how learning in these areas seems to work best. One can't master everything relevant to so broad a field.

Fields of Research

Nanotechnology will mean complete control of the structure of matter, building complex objects with molecular precision. It doesn't exist yet, because we don't have molecular assemblers yet. Work related to nanotechnology accordingly falls into two broad areas: the study of nanotechnology itself (which must remain theoretical, for the time being) and research on enabling technologies leading toward assemblers and nanotechnology (which can be theoretical in part, but which also has an experimental, developmental component).

The theoretical study of nanotechnology involves exploratory engineering work in any of several of areas. It includes basic studies in nanomechanical engineering (the study of molecular machines) and nanoelectrical engineering (the study of molecular and atomically-precise nanometer scale electronic systems). It also includes studies of complex systems, such as assemblers, replicators and nanocomputers. More broadly, it includes studies of non-nanoscale applications, such as large systems built by teams of assemblers.

Because we lack the tools to do real nanotechnology today, these theoretical studies amount to building castles in the air. Accordingly, there is little funding for such efforts and frequent skepticism about their value. Nonetheless, such studies can be pursued with intellectual discipline, yielding firm results and a better understanding of our choices as a society. They have been my main focus and have spawned the current interest in nanotechnology-including the interest in giving these theoretical castles hardware foundations.

Studying what can be done with assemblers yields more foresight than it does progress; working to develop assemblers yields more progress than it does foresight. Inevitably, more resources will go into development than into theory, because technology development will yield practical, short- term results on the way to long-term objectives. It makes no practical sense to try to build an assembler today, but it does make sense to build tools today that will make it easier to build assemblers tomorrow. These tools are termed "enabling technologies."

Promising enabling technologies fall into several familiar categories. These include:

- protein engineering (involving efforts to develop techniques for designing molecular devices made of protein),

- general macromolecular engineering (involving efforts to develop techniques for designing and synthesizing molecular devices made of more tractable materials)
- micromanipulation techniques (involving efforts to extend the technology of scanning tunneling and atomic force microscopy to chemical synthesis and then to the construction of molecular devices).

These approaches have differing strengths and weaknesses. Protein engineering can draw on a host of examples and prototypes from nature and can exploit existing self-replicating machines (bacteria) to make products cheaply-a major consideration, where short-term payoffs are concerned.

General macromolecular engineering avoids the major problem with protein engineering (proteins, not having been designed for designability, are hard to design), but at the cost of moving away from natural prototypes and requiring more expensive chemical synthesis techniques for making near-term products (thus reducing the potential market).

Micromanipulation techniques promise to ease design problems by allowing direct construction of molecular objects, but they suffer from higher costs: a chemical reaction typically makes many trillions of molecules at once, while a manipulator would make but one, hence manipulator-made products can be expected to cost trillions of times more, dramatically reducing the potential market.

Also, as of this writing [1988], micromanipulation has not achieved even a single chemically-specific step in molecular synthesis, while chemists have built specific molecules containing thousands of atoms.

All the above areas bear watching and all will be pursued to some extent, regardless of which ultimately proves to have the biggest payoff. Hybrid approaches, combining techniques from several of these areas (*e.g.*, micromanipulation of molecular tools), seem promising. Finally, improved computational modelling of molecular systems is a generic enabling technology, relevant to all these approaches.

Background Fields: Molecular Science and Technology

There are, as yet [1988], no college curricula aimed at preparing students for work in nanotechnology. My own course at Stanford provided at best an overview of the field. Rather than seeking courses (and books and journals) in nanotechnology, one should seek courses (etc.) in the broad field of molecular science and technology.

Unfortunately, there are, as yet, few (if any) schools that treat molecular science and technology as a unified field. (A note to curriculum reformers: developing a programme having this focus makes sense in terms of current science and technology and would provide a natural home for early studies in nanotechnology.)

Students aiming to gain a solid background in areas important to nanotechnology should be prepared to shop around from department to department. The following section lists some of the important topics and some of the departments in which they are frequently taught.

To study science and technology in a serious way, one must have an adequate background in mathematics. Basic calculus is essential and differential equations and linear algebra are widely used. Problems in nanotechnology vary widely in the mathematical sophistication required for their solution.

The study of physical systems is founded on physics. A knowledge of basic classical mechanics and electromagnetism is essential, as is a knowledge of at least the rudiments of quantum mechanics. Anyone aiming to do any sort of sophisticated work in chemistry and molecular machines can benefit from deeper knowledge of quantum mechanics; anyone interested in molecular electronics should make quantum mechanics a chief focus of study.

"Quantum mechanics" is a broad area, however. The quantum mechanics of interest here is not quantum electrodynamics, quantum chromodynamics, or superstring theory, but the garden-variety quantum mechanics of electrons in matter, the sort studied by chemists and solid-state physicists. Both quantum chemistry and solid state physics are topics of great relevance to nanotechnology.

As with mathematics, so with physics: problems in nanotechnology vary widely in the sophistication needed for their solution.

Nanomachines and nanoelectronic devices are often greatly influenced by thermal noise. To understand its effects, one needs knowledge of thermodynamics and of statistical mechanics. Thermodynamics deals with the flow of energy and heat in matter in bulk; its principles constrain all physical systems and its subject matter is regarded as a prerequisite for the study of statistical mechanics, which describes much the same territory in a more detailed, molecular fashion. These topics are often taught in chemistry and physics departments.

Nanotechnology can be viewed as an outgrowth of chemistry, the leading science in the field of molecular devices and molecular manipulation. Anyone planning serious work in nanotechnology should seek at least a basic background in chemistry, focusing on its structural, molecular aspects. Those interested in assemblers and molecular mechanical devices should study organic chemistry and those interested in the chemical-synthesis path to nanotechnology should study synthetic organic chemistry and learn the arts of the chemistry lab.

Many specific fields have special relevance. Chemical kinetics and reaction transition-state theory is of special relevance to assembler theory. Molecular mechanics is fundamental to any sort of molecular machine design. Studies in materials science (often considered closely allied to chemistry) are also of value; materials scientists consider the mechanical behaviour of larger systems of bonded atoms than chemists typically contemplate.

Biology is the leading science in the study of existing molecular machines. Here, biochemistry is central: enzyme reaction mechanisms provide examples of what many nanomachines will need to do; the folding of proteins and the self-assembly of protein systems provide examples of how complex first generation molecular machines may be made. Familiarity with these fields is of considerable importance to anyone interested in enabling technologies.

Although nanotechnologists will need a thorough grounding in relevant scientific principles, nanotechnology is fundamentally a branch of engineering. To work as an engineer, one must learn to think as an engineer and that means studying (and doing) design. Nanosystems will be systems and so the principles of systems engineering apply. Many nanosystems will be mechanical and so the principles of mechanical engineering apply.

Studies in solid mechanics, system dynamics, mechanisms and control theory all are relevant to both nanotechnology and enabling technologies. Engineering departments often teach more specialized topics of relevance to nanotechnology, such as VLSI circuit design (relevant to nanocomputer design) and microfabrication (relevant to possible enabling technologies). The principles of conventional electronic circuit design are applicable to moderately large nanoelectronic systems and the principles of quantum electronics are applicable to the smallest systems.

Software systems will be vital to nanotechnology and to enabling technologies along the way. A basic introduction to computers and software will be of value to anyone in any area of science or technology. Those interested in software related to nanotechnology should pay special attention to numerical simulation methods for molecular mechanical and quantum electronic systems and to the design of programmes for highly parallel computer systems, since this is the direction hardware will be moving in the coming years. Parallel systems will help designers develop nanotechnology and nanocomputers will later be used to build massively parallel (trillion processor and up) computer systems. Finally, if powerful systems are to be useful in molecular design, they will need to be accessible through fast, clean, intuitive interfaces that let designers see and manipulate model molecules.

Levels of Knowledge

"In short, to do good work in nanotechnology, one must master everything relevant to the physics, chemistry and engineering of molecules, from quantum mechanics to advanced software architectures." Fortunately, this isn't true. Of course, the more you know, the better you'll do (within limits-studying mustn't completely displace doing), but one can't master everything relevant to so broad a field.

What one can and should do is try to master some areas and know a lot about the others. Real molecular devices can do many different things: they can vibrate, pull apart, shake apart, deform, transform, photolyse, or pop from state to state-any of these behaviours can occur in a simple mechanical part and any can make it fail. Real physical systems will do something when used and if what they will do is strikingly different from what you think they will do, then the work you're doing may be a waste of time for you and for anyone who listens to you. It's much better to be right about what will work and this means knowing enough to steer clear of potential problems.

It makes sense to think in terms of three levels of knowledge about a field:

1. Knowing what a field is about-knowing what sorts of physical systems and phenomena it deals with and what sorts of questions it asks and answers.
2. Knowing the content of a field in a qualitative sense-having a good feel for what sorts of phenomena can be important in what circumstances and knowing when you need answers from work in that field.
3. Knowing how to get those answers yourself, based on personal mastery of enough of the field's subject matter.

If one has enough knowledge at levels (1) and (2) in enough fields, then one can steer clear of problems in those fields while doing work in a related field where you have knowledge at level (3). And this is a good thing, because knowledge at levels (1) and (2) takes far less time to acquire. But to make proper use of knowledge at levels (1) and (2) requires a harsh discipline: attempt to assume the worst about what you don't know. Don't assume that a poorly-understood physical effect will somehow save your design; do assume (until finding otherwise) that it may utterly ruin it. Without this discipline, you'll become an intellectual hazard. With it, you'll be able to make a real contribution.

Modes of Learning

How can one get this sort of general knowledge of a field? Courses can help, but they tend to focus on mastery of a narrow range of knowledge, rather than familiarity with a wide range of knowledge. One can gain this familiarity by reading magazines and journals that offer broad coverage of science and technology: good choices include Science, Nature, Science News, Scientific American and IEEE Spectrum. Another good tactic is to skim a wide range of books on the new books shelf of a science library, on a regular basis and to do likewise with a wide range of technical journals.

To do all this properly requires the discipline to read what you don't understand-despite the school-induced reflex which says "Oh, no! I don't understand, so I'll fail the test-maybe I should drop this subject!" By reading what you don't understand, you gain a sense of the patterns of the field-the terms and abstract relationships, the kinds of problems being addressed and the kinds of knowledge required to understand more. And this adds up to an important sort of understanding. Later, this familiarity makes it much easier to consult the literature: one knows which disciplines deal with what problems and what one needs to study to gain a deeper understanding. Also, it fills your mind with questions, so that you can later recognize the answers and have your mind seize them more firmly.

For a thorough grounding in a basic field, classes can be excellent. If classes aren't available, textbooks can often serve well, especially if you work many of the problems.

In any evolving, interdisciplinary field, you must learn to learn from books and journals. Learn to use libraries (as horrible as they are, compared to tomorrow's hypertext publishing systems). Learn to read skeptically-it is a rare book or journal that doesn't have a few serious errors and occasionally publications are utter bilge, especially in interdisciplinary fields (which too often lack any discipline at all).

Finally, tackle problems. If you can find a professor doing good, interesting work, consider becoming an apprentice researcher. If not (or in addition), pursue technical problems that interest you. The best way to learn is to seek answers to questions that interest you and there is no other way to make an original contribution.

Learn to criticize ideas, especially your own. Most new ideas are wrong or inadequate. If you don't reject most of your ideas promptly, then you're almost surely fooling yourself and if you also spread them, you're almost surely polluting the intellectual world. But if an idea really seems to stand up under testing, try filling in more details and criticizing it again.

Get criticism from others. Learn to present ideas in discussions, papers and talks and listen to the responses, especially from people who know relevant fields. If they disbelieve your idea and tell you why, either understand and refute their criticism, or consider working on a different idea. If they look at you oddly and change the subject, consider whether you are perhaps overlooking a really big, basic problem-are you really familiar with the relevant fields? If they disbelieve you at first, but can be persuaded, congratulations! You've probably got hold of something interesting, perhaps even new and important.

Always remember that ideas about real systems must somehow be disciplined by reality. Experimental work brings its own discipline from nature, if the experimenter uses good technique. This discipline is direct and hard to escape. Theoretical work, in contrast, must be disciplined by knowledge of experimental results and natural law; this discipline doesn't impose itself, it must be sought out and largely self-applied. To be a careful thinker, try to understand things in more than one way: if you get the same answer from physical calculations and from analogies to known machines and from analogies to biology, then you're probably right. If all you have is a rough analogy or a crude calculation, you may well be wrong.

Seek out weaknesses in ideas and build only on ideas that pass rigorous tests, or you may see the foundations of your thinking later crumble and dump a year's (or a decade's) work into the trash. Beware of those who have neither experimental results nor a theoretician's voluntary discipline; expect them to spout great streams of plausible nonsense, unconstrained by reality. Don't become one of these, even if you find that many (ignorant) people are intrigued and entertained by your wilder imaginings.

In short, learn the fundamentals of molecular science and technology. Survey other relevant knowledge. Learn to learn from books and journals. Pursue problems, think critically and learn more. Design and calculate or experiment. Publish your contribution and add to the world's knowledge.

The Study of Nanotechnology

Students interested in nanotechnology often ask what they should study. This web page provides a partial answer to that question. Foresight has a briefing on the subject by Eric Drexler.

Nanosystems

The standard text in the field is Nanosystems: molecular machinery, manufacturing and computation by K. Eric Drexler. Buy a copy and study it.

Molecular Mechanics

Any manufacturing technology must move atoms from where they are to where we want them to be. How atoms move and the forces that act upon them during their motion are therefore critical areas of study in nanotechnology. This field is called molecular mechanics. A very brief discussion of molecular mechanics and its significance for nanotechnology is available on the web in Computational Nanotechnology. It contains references to further reading.

A classic introduction to molecular mechanics is Molecular Mechanics, by Ulrich Burkert and Norman L. Allinger, ACS Monograph 177, American Chemical Society, 1982. While out of print, it can often be found in campus libraries.

Nanosystems introduces the basic concepts of molecular mechanics in the book. A great advantage of Drexler's treatment is the adoption of consistent SI units. A slow and careful reading of this book is worth the effort.

Many other introductions to molecular mechanics are available. Software packages implementing specific approaches to molecular mechanics are available and can be very useful in learning the concepts.

Positional Control, Stiffness and Elasticity

A central idea in nanotechnology is that of positional control. This can be provided by fairly standard robotic devices. A major difference between conventional robotic devices and the molecular robotic devices used to position molecular components is the need to consider thermal noise. At the molecular scale, components wiggle and jiggle because of Brownian motion. To control this, the component must be held stiffly, *i.e.*, there must be a restoring force which acts to return the component to an equilibrium position if it is displaced. (The existence of a restoring force serves as a good abstract definition of "positional control.") The restoring force is usually assumed to be linear in the displacement: restoring force = ks times displacement. The constant ks is a measure of the stiffness of the system. The greater the stiffness, the greater the restoring force and the smaller the deviations of the system from its equilibrium position. The basic equation relating stiffness and positional uncertainty is:

sigma2 = kT / ks

This is equation introduced in the book of Nanosystems. This equation should be memorized and basic applications understood. To use it, it is essential to determine the stiffness ks. The stiffness of a structure can be determined from its geometry and material properties. The basic concepts are introduced by Feynman in book of The Feynman Lectures on Physics, by R. P. Feynman, R. B. Leighton and M. Sands, M. Addison-Wesley (1964). Buying the Feynman Lectures is recommended. Read and understand the book.

Application of these equations to some robotic devices (including the Stewart platform which, because of its great stiffness, is very attractive for molecular robotic applications) is illustrated in A New Family of Six Degree of Freedom Positional Devices. Applications are also discussed in the book of Nanosystems and in section 13.4 which discusses a robotic arm.

Self Replication

A second central idea in nanotechnology is that of self replication. The student should read the web page introducing self replication and select some of the references therein for further reading. The recursion theorem is basic to self replicating systems. This theorem should be understood. Exercise: write a programme which prints out an exact copy of itself. Buy a copy of Kinematic Self-Replicating Machines, it is an excellent survey of the literature.

Further Study

There are, of course, a great many other subjects relevant to the development of nanotechnology. It seemed more useful to provide a short and focused list of critical subject areas that could be mastered with a reasonable effort than a longer and more unwieldy list which included everything of importance. The student can rest assured that there is no shortage of material to study relevant to this new field of research.

A few relevant academic institutions:

- Caltech's Molecular Simulation Centre
- USC's Laboratory for Molecular Robotics

Exploring Career Options

What is Nanotechnology?

Nanotechnology is a way overused term and I think the better term is nanoscience. Nanoscience is the study of things that are as small as molecules. Small things act very differently from large ones. For example, a gold nanowire is twenty times stronger than a large bar of gold. Electrons move differently through nanowires than through regular wires. We still have a lot to learn about the fundamental properties of nanoscale structures.

On the other hand, nanotechnology is about trying to make technologies, such as computers and medical devices, out of these nanoscale structures. Although the field of nanotechnology is fairly new, one might say that nanotechnology has been around for a while. The zinc oxide nanoparticles in that white stuff that people put on their noses for sunscreen could be considered a nanotechnology, for example.

In my lab, we do both nanoscience and nanotechnology. We work on designing and making the building blocks of nanotechnology: nanowires. We also are studying the fundamental properties of these building blocks. Finally, we are developing ways of organizing them into larger structures that could ultimately become part of useful technologies, such as computers.

How did you become Interested in Nanoscience?

After graduate school, I studied the behaviour of materials that were so flat that they were essentially two-dimensional, such as the sheets you can peel off from mica and graphite.

That led me to want to make materials that were one-dimensional-very skinny and long, like wires. So I started to work on the problem of developing a nanoscale wire. I also realized that these wires could be used to move information around in nanoscale devices, like computers.

What can be done with Nanowires?

In the future, we will be able to make many common technologies using nanowires, but the processes we will use to make them will be very different from how those same technologies are manufactured today.

For example, transistors are crucial elements in computers. Today, manufacturers etch them out of silicon chips using multi-billion-dollar fabrication lines. The process is very expensive.

In nanoscience, however, you can build from the bottom-up. You can literally take a bottle that contains a nanoscale silicon wire in a fluid, pour the contents onto some kind of surface and make a nanoscale transistor. It's cheap. My lab is also exploring ways of using interactions between chemicals or proteins to assemble nanoscale devices.

Of course, the next step is learning how to organize these small structures into larger devices, from computers to lasers to biological sensors.

What is a Biological Sensor?

Most biosensing today requires various chemical steps. When you get your blood drawn to get a test done, clinicians go through various procedures to see if you have some virus or protein present in your blood that is an indicator of disease. It can take hours or days.

On the other hand, our biological sensors use these little wires, which behave as transistors that turn on and off in response to specific biological molecules. The way transistors turn on and off in a computer is that you apply a voltage to them. But at the nanometer scale, that's the same thing as binding a charged molecule to a wire. Proteins and viruses are charged, so we get a signal from the transistor when a protein or virus binds to the surface of the wire. The wires are treated to accept only the proteins or viruses we're interested in, so we know immediately if the protein or virus is present in the blood.

How could a Biological Sensor be Used?

Right now, someone who's diabetic can buy a device that allows them to get their glucose level from just a pinprick of blood. Similarly, with the biological sensor, you could buy a device at CVS, spit on it or take a little pinprick of blood and determine if you have the flu or any other viral disease.

I know from talking to doctors from the National Cancer Institute that they don't have a way to do real-time monitoring of chemotherapy, so the doses they give patients are fairly crudely calculated. But if you could actually use a biological sensor to monitor the cancer marker proteins as the chemotherapy is being done, you could provide a much more sophisticated level of treatment for cancer patients.

We also have some really interesting but unpublished work for the defense department relating to biowarfare and chemical warfare agents. Most of the devices out there today for detecting these agents are one-element sensors-that is, they can only detect one type of molecule at a time. Like many blood tests, they're not very fast. But in my lab we can routinely make sensors with ten elements that can simultaneously sense ten different toxins. You can get the readout in real time of whether there's cholera, botulin toxin, or other things around. I think it's a quantum leap ahead of any other technology, especially for real-time monitoring.

The sensor is a good technology because you don't need it to be very complex for it to be useful-a 10-element sensor is 10 times better than any other test you can do today. A hundred elements are almost unimaginable to a doctor or clinician. The sensor is an application that can have a positive impact on society, while at the same time serving as a stepping stone to more complex integrated systems.

How Long before Biological Sensors are Widely Available in Medicine?

The science is almost done and all of the pieces are essentially there. The major limitation at this point is not technical, it's economic. A lot of venture capitalists don't feel that diagnostics is a sufficiently profitable area to pursue. It's also very difficult for any new technology to break into the market.

I'm very naive in business. I hope that if something can really help people, we can find a way to make it available for them. It's just going to take a lot of pushing on my part and others'.

How Soon do you Think that we will be Able to Pick Something out in the Nanotechnology Section of Best Buy?

I would say 10 years. It's going to be a while before we make nanotech electronic devices that are as sophisticated as electronics are today and Best Buy won't be interested in selling them until they are. But other markets might be more approachable.

For example, maybe you'll go to the clothing store to buy nanotechnology. Something we're working on is making devices in completely flexible plastics so you can have ubiquitous computing. You could wear a T-shirt that has your Palm Pilot built into it.

I believe that we're still waiting for the "killer app" for nanotechnology. I think we need to think about the applications that best take advantage of the unique characteristics of nanoscience. What unique thing can we make from this technology that people really need? Once we figure that out, nanotechnology could be available in just about any store.

What's the Most Exciting Challenge in Nanoscience that a young Person might tackle in the Next 10 or 20 Years?

One of the biggest challenges is developing the underlying science and technology of how to assemble and integrate nanoscale structures, using our own bodies as an analogy. We start out from one simple cell, which divides and ultimately builds this really complex organism. Similarly, a single nanoscale device could be the starting point for making a computer if we could encode it with the right information. Once we know how to do that, we could build just about anything we can imagine. The person or people who solve that problem are going to completely open this up as a technology.

Another challenge for the future is to build at the interface between biology and man-made devices. If you take our biological sensors as the most primitive example, the sensor is actually converting a biochemical signal into a digital signal. It suggests that we have a way of building a very natural interface between biological and digital systems.

From what kinds of Scientific Backgrounds can you Approach Nanoscience?

Essentially any science. People who work with me come from biology, chemistry, physics and engineering backgrounds. I haven't stayed at all in chemistry-nanotechnology has given me an opportunity to continually learn new things. People who are very broadly knowledgeable are the ones who can put together the ideas that make the new steps forward. The field is completely interdisciplinary-you can't stay focused in a traditional core area and make the breakthroughs that will drive the field in the future.

A lot is made of ethical issues in nanotechnology. Do you worry about making self-propagating devices that will run amok?

I think the concern about self-propagating machinery is not an issue right now in science, except possibly in re-engineering viruses, which are already adapted to take over cells. We don't re-engineer viruses in my lab. In nanoscience, one needs to take precautions more in the same way that you would take precautions with chemicals: You don't want to ingest radioactive chemicals, carcinogens, or toxins.

Do you have any Words of Advice for our Readers?

For young people interested in science or engineering, nanoscience is a fantastic area. There's so much going on, it's changing really fast and it's clear that we're waiting for someone really innovative to come and lead the field in a better direction. I'm hoping I can do that in part, but I think there's a real opportunity for a new generation of scientists.

Careers in Nanotechnology

The progress of nanotechnology requires the interaction of many scientific and business skill-sets from biology, chemistry and engineering to entrepreneurship, business and intellectual property management. The creation of human and material infrastructure for this progress is therefore largely dependent upon the commitment to skill development within the education sector and within the research arms of major industry.

Skills: Nanotechnology draws upon several disciplines, chief of which are:

- Physics and chemistry, particularly materials and surface sciences
- Biology and biotechnology
- Engineering

A broad-based science degree, with an emphasis on materials science and molecular biology, provides an ideal platform. However, nanotechnology will provide opportunities for scientists and engineers with most combinations of scientific and engineering disciplines.

Courses and Nanotechnology Studies

Australia led the world with the development of the nanotechnology degree course via Professor Joe Shapter of

Flinders University in 2003. In 2004 RMIT University introduced a Nanotechnology Masters degree course. All major universities in Victoria have nanotechnology research projects underway as part of Applied Science or other scientific study programmes and nanotechnology is now being considered for integration into secondary school curricula throughout Australia and New Zealand.

Tertiary Courses: There are a significant and growing number of tertiary courses being offered in nanotechnology across Australia:

- Nearly all universities in Australia offer nano-relevant masters or PhD studies in physics, chemistry, mathematics and engineering.
- Undergraduate nanotechnology degree courses are offered at: University of New South Wales, University of Technology Sydney, University of Sydney, Curtin University, RMIT University, University of Western Sydney, La Trobe University, Flinders University and University of Wollongong (Medicinal Chemistry). These are typically 3-4 years in duration with either a chemistry or physics major and at least one cross-curricular nanotechnology subject each semester.
- Deakin University will offer a nanotechnology degree from 2006.

The Science Show on ABC Radio National ran an interview on Nanotechnology: the Science of the Future on Saturday 19 March 2005 discussing the undergraduate degree courses in nanotechnology being offered by several Australian universities. Click here for the transcript.

Nanotechnology Victoria is leading the way in developing an integrated approach for Victorian nanotechnology, based initially upon the educational activities of its Members. Each of Monash, Swinburne, RMIT and the CSIRO play a role in the development of scientists, engineers, technicians and other professional disciplines.

Career Opportunities: The broad range of skills required for nanotechnology also lends itself to a number of possible nanotechnology careers. These include:

- research and development in industry and government, including health, environment and materials;
- education and research in universities;
- product development and commercialization;
- marketing and communication of new products and new technologies;
- management and entrepreneurship for start-up companies developing nanotechnologies;
- investment advice in regards to new industries and technologies.

Nanotechnology Careers

What is the Workforce Need? **:** As nanoscale science and technology come to have increasing impacts on many aspects of our daily lives, the opportunities for careers in these fields are expanding rapidly. A major challenge for the field is the education and training of a new generation of skilled workers.

Nanotechnology job projections are estimated to be nearly two million workers worldwide by 2015. In what countries will these jobs occur?

- 0.8-0.9 million—USA
- 0.5-0.6 million—Japan
- 0.3-0.4 million—Europe
- 0,2 million—Asia Pacific (excluding Japan)
- 0.1 million—other regions

In addition, nanotechnology will create another five million jobs worldwide in support fields and industries.

Where are the Career Areas?: In areas as diverse as designing medical diagnostic devices to building better batteries, from creating cosmetics to enhancing energy efficient windows, from auto and plane manufacturing to researching the nature of matter itself, knowledge of nanoscale science and technology will be increasingly important during upcoming years and decades.

Current applications of nanoscale science and technology and thus career opportunities, exist in areas such as:

- Electronics/semiconductor industry

- Materials science including textiles, polymers, packaging, among other
- Auto and aerospace industries
- Sports equipment
- Pharmaceuticals including drug delivery, cosmetics, among others
- Biotechnology
- Medical fields
- Optoelectronics
- Environmental monitoring and control
- Food science including quality control and packaging
- Forensics
- University and federal lab research
- National security
- Military
- And many more

Nanoscale science and technology are fuelling a revolution in manufacturing and production, creating new materials and novel processes. Not only will the areas listed above continue to grow and benefit from nanotechnology, but the following fields are expected to undergo explosive developments:

- Medicine: diagnostics and therapeutics (*e.g.*, drug delivery)
- Energy: capture, storage, & use; fuel cells, batteries
- Environmental remediation: in conjunction with GM microbes
- Robotics: many uses
- Manufacturing: self-assembly; "bottom-up" fabrication of novel materials
- Commerce: Radio Frequency Identification (RFID) "smart" tags
- Space exploration: space elevator

As these lists of nanoscience-based applications indicate, our world is increasingly dependent on science for food, shelter, energy, etc. For our democratic society to function effectively, citizens must become familiar with at least some basic science and, perhaps even more importantly, with thinking scientifically.

What Education is Needed and in Which Fields?: Nanoscale phenomena underlie many of the properties and interactions of matter and thus the sciences of physics, chemistry and biology. Studying these fields and paying attention to the developments in nanoscience that advance them and the applications in nanotechnology that they support, can provide you with a solid foundation for any of a broad range of careers.

Potential fields of study include:

- Biology
- Chemistry
- Physics
- Environmental Science
- Agricultural Science
- Engineering
- Medicine
- Forensic Science
- Law
- Business
- Ethics

What Type of Education is Needed for a Career in Nanotechnology?: Not everyone working in the field will require a doctorate degree in one of the fields noted above. A skilled workforce trained at a variety of levels is needed to meet the projected workforce challenge of 7 million workers. Education levels include:

- Technical programme
- Associates degree (2 years)**
- Bachelors degree (4 years)**
- Masters degree (6 years)**
- Doctorate (8 years)**

** This is the estimated time it would take a full time student to complete degree requirements after graduating from high school.

What are the Expected Salaries?

- Two Year Associates $30,000-50,000
- Four Year Bachelors $35,000-65,000
- Six Year Masters $40,000-80,000
- Eight Year Doctorate $75,000-100,000

Source: Pennsylvania State University; Centre for Nanotechnology Education and Utilization

Nanotechnology Jobs

This site is dedicated to nanotechnology jobs in small tech, including careers in nanotechnology, micro-technology, biotechnology and information technology. Here you will find nanotechnology employment in such disciplines as chemistry, physics, materials science, biology, biochemistry, molecular biology, biomedical engineering and devices, microfluidics, microarrays, information technology, optics, mechanical, electrical and chemical engineering, micro- and nano-electromechanical engineering (MEMS and NEMS) and other relevant fields.

Nanotechnology Job Seekers: In addition to information about working in nanotechnology, job seekers can search a database of nanotechnology jobs, find out about nanotechnology employment opportunities at top companies and find a wealth of educational resources to further their careers in nanotechnology. Other resources on the site include updated industry news, an events calendar and a database of associations and societies.

Nanotechnology Employers: Tinytechjobs makes it easy for employers to post jobs and search for resumes. Job postings are only $195 each, stay up for 60 days and reach a highly targeted audience across our network of 16 related sites. And, now you can purchase job posting packages and resume database access online in minutes! You must register to post jobs and search resumes. Registration is FREE and takes only minutes.

Advancing Beneficial Nanotechnology

Foresight is the leading think tank and public interest institute on nanotechnology. Founded in 1986, Foresight was the first organization to educate society about the benefits and risks of nanotechnology. At that time, nanotechnology was a little-known concept.

Today, with the basic framework of public understanding in place, we are refocusing our efforts on guiding nanotechnology research, public policy and education to address the critical challenges facing humanity.

Foresight's new mission is to ensure the beneficial implementation of nanotechnology.

Foresight is accomplishing this by providing balanced, accurate and timely information to help society understand and utilize nanotechnology through public policy activities, publications, guidelines, networking events, tutorials, conferences, roadmaps and prizes.

Foresight is a member-supported organization. Our membership, including over 14,000 individuals and a growing number of corporations, is diverse demographically and geographically. They are interested in ensuring that the future of nanotechnology unfolds for the benefit of all. These concerned individuals include scientists, engineers, business people, investors, publishers, artists, ethicists, policy makers, interested laypersons and students from grammar school to graduate level.

Careers in Nanotechnology

Nanotechnology represents a recent area of specialisation and graduates follow a diverse range of pathways following training in this exciting new field. Examples of careers available for graduates include:

- entrepreneurial, management and investment advisors in biotechnology and research and development industry
- research and development in government, universities and private research institutes
- education and academia
- health industry research and consulting—pharmaceutical, medical, agriculture, food and beverage, environmental industries

- product development and advising
- communication and media, marketing of new technologies
- many new industries emerging as a product of advances in nanotechnology

Careers in Biomolecular Nanotechnology

The beginning of the 21st century marks a critical juncture in technology, as the size scale of electronic elements has decreased to the hundred nanometer range and the complexity and size of the chemical and biochemical structures we can design and manipulate has increased to a similar size range, a size which also matches the wavelength of visible or UV light more.

This has opened up a whole new field of possibilities ranging from molecular electronics (using molecules themselves as circuit elements) to direct bioelectronic interfacing (communicating with and/or controlling biological systems with electronics or optoelectronics).

Humankind is rapidly approaching a point where we can manipulate our living environment well enough and build complex electrical, optical or mechanical devices with enough complexity so that it will become possible to interface or even fuse biology and technology at a fundamental level. In medicine, this could mean real-time monitoring of bodily functions, replacement parts (biological, non-biological and hybrid), "smart" drugs which can be controlled externally and ultimately entirely new physiological functions such as enhanced sensory perception (for example, built-in sensor arrays) or the ability to control external devices by thought, mood or activity level.

In computing, this means electronic elements which are themselves molecules, not only offering extraordinary element density and speed, but enhanced possibilities for direct interfacing of computers with biological systems. In the environment or in agriculture, this could mean designing new species of plants and animals that are directly or indirectly interfaced with computers and that monitor information, process it and alter their environment accordingly (or signal other devices which alter it).

From this description, it is clear that a student in this field needs to learn more than just the lab science and engineering required to perform the manipulations. Science and Engineering training can no longer effectively be performed in isolation. Students need to consider:

- How will Biomolecular Nanotechnology interface with commercial enterprise and what opportunities exist there?

Student Internships

The interface between Biomolecular Nanotechnology and commercial enterprise. This is a fine academic concern, but are any real companies interested in developing this kind of technology? Absolutely. We have developed a series of collaborative interactions with companies that involve joint research and student internships. These companies are not only interested in developing the technology, they are interested in the possibility of hiring our students. What companies? How about:

- The Biochip division of Motorola which makes DNA biochips for sensing specific DNA sequences hybridizing to large arrays of DNA probes.
- The Digital DNA division of Motorola which is pushing forward in microfluidics (chemical laboratory on a chip methodologies) and molecular electronics.
- The Reconnaissance division of Lockheed/Martin who is interested in living biosensors.
- Kodak who is interested in molecules that respond to and communicate with light.
- QTL a recently founded company working on smart materials for sensing applications.
- Molecular Imaging a relatively new company which develops Scanning Probe Microscopy for detecting, imaging, characterizing and communicating with molecules on a surface.

These companies are directing involved in our training programme, coming to ASU and teaching our students about corporate structure, intellectual property, company research and entrepreneurial ventures. In addition, they have asked to have interested students come and work with them and learn first hand about how this new technology is being developed at the industrial level.

- What is happening in the broader fields of bioengineering, molecular electronics, biomaterials, biophysics and molecular biology?

Attending Biomolecular Nanotechnology Meetings

One of the most important activities outside of ASU that we encourage all of our students to engage in is active participation in national and international scientific conferences. All or part of the expenses for such meetings (depending a bit on the circumstances-sometimes the cost is shared with the graduate programme or advisor) will be paid for by one of the general funding sources available at ASU for study in Biomolecular Nanotechnology. Attending meetings serves several purposes.

The first is obviously education in areas beyond those covered by ASU faculty and exposure to viewpoints different than those held by ASU faculty. Second is exposure of the student's work to the scientific community and establishment of the student as a nationally or internationally recognized scientist. Finally, conferences are where students meet potential future postdoctoral advisors or industrial employers.

- What is the rest of the world doing in the area of Biomolecular Nanotechnology?.

Visiting Nanotechnology Laboratories Abroad

The revolution in nanotechnology and its biological implications is not confined to the US. Researchers particularly in Europe and Japan have made large strides in this area particularly as it concerns the optical communication with individual molecules and the development of molecules with new and interesting electronic or mechanical properties. We have collaborators around the globe involved in these efforts. One of the best ways for a student to increase his or her awareness of the different kinds of research taking place in the world is by going to work for periods of a few months in a laboratory in another country. In addition, science culture, like societal culture, varies significantly from place to place, with each system having its strong and weak points. Students will have the opportunity to participate in international exchanges, funded by the training programme.

- What are the social and ethical impacts of this new technology and how will our government control and fund it?

The Citizen Scientist

Humankind is approaching a point where we are going to be able to control our own evolution and that of the living world around us. We will have the option of modifying ourselves and our world both through re-engineering of present functions and the development of entirely new functions, some of which may be synthetic.

The potential impact of this technology on our culture and on cultures around the world is enormous. Given the small number of citizens well trained in these areas, our students will need to be able to lead society in discussions of the ramifications of new technology, both providing information and gauging societal response. In addition, they will need to be able to help shape the science policy that will dictate further technological development. In addition, the way in which companies and universities deal with intellectual property and interactions with the media has become an increasingly important issue.

To address these needs, students in our programme will explore the political and societal mechanisms by which decisions concerning science funding and policy are made and what impacts these decisions have had on society in the past. This will begin with the second semester course at ASU and culminate with a two-week long workshop in Washington, DC, which will expose graduate students to a broad range of policy issues related to research and development.

The ASU-based training consists of a one semester course in which students will interact with our industrial affiliates, including large (*e.g.*, Motorola) and small (*e.g.* University start-ups) companies, learning how Ph.D. level science and engineering works in corporate America, how an entrepreneur takes an idea from concept to start-up company and what is involved in the protection of intellectual property. We will also discuss government funding of research and begin to consider the societal impact both of research and research funding decisions in the nanotechnology area.

The Washington DC—based training module will engage the students in a series of seminars and exercises, with the goal of allowing them to better understand the complex interactions among science, technology and democratic society. The module will be designed and implemented by Arizona State University's Centre for Science, Policy and Outcomes.

The formal component of the module will be a series of half-day seminars presented by leading thinkers and practitioners in the areas of science policy and science and technology studies. Speakers will primarily be drawn from federal science agencies, congressional staff and Washington area universities and think tanks, with a few speakers brought in from out of town. Seminars, which will be strongly rooted in real-world cases, will cover such issues as:

The federal R&D policy process

The rationale for publicly funded science

Issues and problems in human resources for science and technology

Science, technology and justice

Expertise and policy making

Ethical issues in science and technology

Science and democracy

International issues in science policy

Science and the press

In addition, students will engage in several formal and informal exercises, including visits to congressional offices, agencies and non-governmental organizations; attendance at hearings, press briefings and other public fora; mock interviews;

and press releases; guided assessments of museum exhibits; and group discussions about significant issues relating to science and society. The module will culminate with a participatory exercise such as a mock congressional hearing or press briefing.

A key objective will be to help students appreciate that questions that may seem simple and strictly "scientific" in the laboratory may in fact involve complex questions of values and choice in the real world. We hope to heighten students' appreciation for and insight into the democratic processes by which such questions are resolved and to challenge them to think about their own research in a broader societal context.

Master of Science in Nanoscale Physics

The Nanoscale Physics programme prepares students for a career in nanoscience by combining a strong component in quantum theory, which governs the behaviour of systems at the nanoscale, with the study of practical nano- and mesoscale devices. This provides the student with the knowledge required to successfully navigate the emerging field of nanoscale science and nanotechnology. In addition, a year-long course in methods of experimental physics is offered to ensure that students obtain the advanced practical skills valuable to the nanotechnology industry.

Rice is a well-established centre for nanotechnology, with researchers active in several departments outside the physics core. The Richard E. Smalley Institute for Nanoscale Science and Technology includes among its faculty Nobel Laureate R. Curl, as well as esteemed faculty from the physics and astronomy, chemistry, electrical and computer engineering, civil and environmental engineering, chemical engineering, bioengineering, computational and applied math and mechanical engineering and materials science departments. It focuses on interdisciplinary studies in many areas of nanoscale science including carbon nanotubes, nanoshells and nanobiology.

The Centre for Biological and Environmental Nanoscience (CBEN) at Rice fosters the development of this field through an integrated set of programmes that aim to address the scientific, technological, environmental, human resource, commercialization and societal barriers that hinder the transition from nanoscience to nanotechnology. The Centre's research focuses on investigating and developing nanoscience at the "wet/dry" interface.

What it Takes to Work in Nanotech?

If scientific disciplines were dishes on the menu of a Chinese restaurant, nanotechnology would be accompanied by five red stars-the field is hot. It should come as no surprise then, that along with headlines and boatloads of government funding and investment dollars, nanotechnology is also grabbing the attention of jobseekers. But what does it take to work in nanotechnology? ACS's Office of Career Development asked four scientists working at the forefront of the field to tackle that question during a presidential event at the society's national meeting in Anaheim, Calif., late last month.

Because nanoscience is such a young field, the panelists didn't have a well-trodden path to follow to their current positions. While each took a different, often circuitous route, they were all able to agree that there are certain skills that are key to a successful career in nanotech and also a successful scientific career in general.

Harris A. Goldberg takes a can of tennis balls almost everywhere he goes. It's pride and not some relentless drive to improve his backhand, that leads Goldberg, the president and CEO of InMat, to pick such a peculiar accessory. Tennis balls, specifically Wilson's Double Core tennis balls, have made Goldberg and InMat among the most well-known purveyors of nanotechnology-enabled materials. That's because the Double Cores employ InMat's nanocomposite-based Air D-Fense coating to maintain bounce while also retaining air.

When describing the early days of InMat, Goldberg cannot help but talk about risk. Trained as a physicist, Goldberg spent his pre-InMat years weathering corporate restructuring while working on nanocomposite barrier coatings for big chemical companies like Hoechst and DuPont. Eventually, he says, the time came when he could see that his project, one he strongly believed in, was going to be cut back. So he asked his team if they were ready to take a leap of faith and enter the risky world of start-ups.

Goldberg says it's important to understand the risks and the likelihood of success before joining or starting a small company. He explains that it took years of inspiration, perspiration, desperation and serendipity before his risk paid off. He and his partner, Carrie Feeney, initially used their own savings to pay InMat's employees. For three years, Goldberg and Feeney worked without pay, relying on their spouses to support them.

For scientists looking to start their own nanotech companies, Goldberg recommends finding a new nano-enabled technology for an established market. That way, you're not trying to make some totally new product- nanoelectromechanical devices, for example -that may not have a market and you're also not competing with dozens of other companies that have established nanotech research programmes.

But Goldberg adds that the skills required to start a successful nanotech company—choosing the right business model, assessing technical risks, hard work and the ability to take risks—are the same skills needed to start most technical businesses.

When asked if a career in nanotechnology is different from any other career, Margaret Blohm, manager of General Electric's nanotechnology AT programme, gives three different answers: yes, no and maybe—all of which, she says, can be the right answer simultaneously.

Nanotechnology is a new field, so yes, she says, it requires skills and knowledge that don't necessarily come with a traditional scientific education. But then again, Blohm says, a career is a career, so no, the skills that make someone successful in nanotechnology are really no different from the skills that make someone successful in any other scientific field. And finally, because nanotechnology is really an enabling technology that will ultimately have to work within different disciplines, maybe what a scientist defines as "a career in nanotechnology" simply comes down to semantics.

Originally Trained as an organometallic chemist, Blohm has spent the past 15 years working for GE. While most of her work has been with polymers and plastics, she says that she saw the chance to manage nanotechnology at GE as "the opportunity of a lifetime."

Working in nanotechnology, Blohm says, meant learning to be comfortable with being different from everyone else. With nano, everything is new, she explains, but in a world where everyone is doing things a more traditional way. Often, she finds she has to convince people why doing things the new way can sometimes be better than the conventional way. The feeling can be unnerving. But Blohm says that because she wasn't afraid to be different, her alternative approach turned out to be an advantage. "My geek years in high school have paid off," she jokes.

Although it may sound like a cliche, Blohm also emphasizes that when it comes to nanotechnology, being able to work in a multidisciplinary team is essential. As a manager of a group that includes chemists, biologists, physicists and engineers, she says that she often feels like she's herding cats, adding, "But I like cats." Ultimately, Blohm says, a group's ability to collaborate and to have constructive disagreements is what moves a project forward.

Rather than seeking out employees with particular skills in nanotechnology, Blohm says that GE aims to hire the best scientists that it can and let them learn on the job. The company looks for an impressive publication record and other demonstrations of expertise. And while she stresses that scientists should be passionate about and should possess a depth of knowledge in their area of expertise, she adds that they should also be curious about other areas of science. Beyond that, Blohm advises that having fun should be a priority in any job.

It would be easy to look at Charles Z. Hotz's job history and imagine that it was all part of a grand plan culminating in his current position as director of R&D at Hayward, Calif.-based Quantum Dot. But to be truthful, Hotz says, his career path was far less calculated than it seems.

Although he earned his Ph.D. in physical organic chemistry, Hotz worked in myriad disciplines- polymers, materials science, DNA synthesis and sequencing, fluorescence detection and biolabeling -before landing his current job leading R&D efforts at Quantum Dot, a company that makes nanocrystals with a multitude of applications.

Hotz says that his desire to learn has been the driving force behind each of his career moves. With each new job has come new challenges but also new skills. While he jokes that he's "had the benefit of many graduate educations" during the course of his industrial career, he adds that he approached each of those jobs as a chemist. And he couldn't have been successful doing it any other way. When it comes to career advice, Hotz prefers to speak in general terms rather than focus on one or two nano-specific skills. "Career management is an important part of a successful career," he says, "and a career in nanotechnology poses few unique alterations to it."

Hotz advises chemists to keep learning at all stages in their careers. "If you're not acquiring new skills as time goes on, you're not becoming more versatile or more marketable."

For industrial research chemists who are further along in their careers, he emphasizes the importance of learning to communicate beyond the R&D sphere. "If you want to become more influential, you need to influence people who are outside of R&D, even if you're working in R&D." That means learning to effectively communicate with people in the operations, marketing and legal departments of a particular company.

The advice is especially important for scientists looking to move into leadership roles. "You'll never move up beyond R&D if you don't understand what's beyond R&D," he says.

Hotz tells chemists in the early stages of their careers not to myopically focus on one particular specialty, especially during graduate school. It's too hard to predict the direction that chemistry will take in the next 10 years, he argues. A chemist is best served by a strong grounding in a fundamental field.

"If you really work hard to be a great chemist, then you'll be a great nanotechnologist," Hotz explains, "because chemistry is always going to be critical in nanotechnology." Or to put it more succintly: "Be a good chemist. Don't worry about nano."

James C. Romine says that he began his career with DuPont because it was the only company to offer him a job that wasn't related to his graduate work in the liquefaction of coal. Like Hotz, Romine's winding career path has taken him in many directions, although all of them have been under the aegis of DuPont.

During his 24 years with the company, he has worked as both a research scientist and a manager in plant operations, biopolymers, high-temperature materials and lycra and nylon, until reaching his current post as director of DuPont's materials science and engineering department. Romine concedes that it's not a route he expected to take back when he was in graduate school studying physical organic chemistry.

He says that if there's one thing he's learned while managing his career, it's that "you've got to sell yourself in everything you do." Romine asks: "What do you want to do? Can you make a difference? Are you a cog or a leader?" Selling yourself, he explains, means being able to effectively answer these questions.

Selling yourself also means showing that not only are you flexible as a scientist, but you also have broad interests beyond your specializations. To illustrate the last point, Romine lists the thesis titles of some of the chemists that DuPont recently hired to work in nanotechnology. He points out that not all of the titles include words with the "nano" prefix. But all of them were interdisciplinary projects.

Romine also says that DuPont is most interested in chemists with doctoral degrees. He sees the degree as evidence of both tenacity and a passion for science. "A Ph.D. gives you the ability to fail nine out of 10 times," he says half-jokingly. "It means that you can get up every day and be excited about things you know almost certainly won't work."

When interviewing prospective employees, Romine says that he asks at what age they decided to become a chemist. "If it's under the age of 10," he says, "I hire them." Of course, that doesn't mean he won't hire a candidate who decided to be a chemist later in life. It's just that he thinks that anyone who made a career decision at the age of 10 is following a dream rather than making a career choice based on external factors like salary and employment prospects. People who are following their dream, he says, "are the people who are the most excited about the things they do."

In talking with students who are interested in nanotechnology, Romine finds that many are too focused on learning a particular skill or getting into a certain programme. He tells them, "Go where the problems are harder. Because if you solve a hard problem, people will notice. And when people notice, that's what makes careers."

Atomically-precise Nanotechnology

Palo Alto, Calif., May 8, 2007—Foresight Nanotech Institute, in partnership with the Society of Manufacturing Engineers (SME) and with the support of Battelle, announced today that it will team up to unveil the ground-breaking Technology Roadmap for Productive Nanosystems at a new nanotechnology event, the Productive Nanosystems Conference, which will take place October 9-10, 2007 at the DoubleTree Crystal City in Arlington, VA.

In 2005, Foresight Nanotech Institute, a leading nanotechnology think tank and public interest organization and Battelle, a leading global research and development organization, launched development of the Technology Roadmap for Productive Nanosystems through an initial grant from The Waitt Family Foundation. The group assembled a world-class Steering Committee to guide this ground-breaking project and garnered the support of several important industry organizations as roadmap partners, including SME. The Productive Nanosystems Conference will launch the first version of this new nanotechnology Roadmap.

Productive Nanosystems are functional systems that make atomically precise structures, components and devices under programmable control. This is driving research and applications in a host of areas, with the goal of providing new atomically-precise nanoscale building blocks, components and devices.

The aerospace, defense, medical, electronic, building/construction, automotive and many other industries will benefit from productive nanosystems. In the aerospace world, productive nanosystems will be essential for production of new, lighter weight, stronger, temperature resistant materials. In the medical industry, productive nanosystems will be key to many advanced medical diagnostic and therapeutic applications. In the automotive field, productive nanosystems will become central to the development of high performance materials, electronics and information technology for Tier 1 suppliers.

About Foresight Nanotech Institute: Foresight Nanotech Institute is a leading think tank and public interest organization focused on nanotechnology. Founded in 1986, its mission is to ensure the beneficial implementation of nanotechnology. Focusing on the six Foresight Nanotechnology Challenges, Foresight provides balanced, accurate and timely information to help society understand nanotechnology through publications, public policy activities, roadmaps, prizes and conferences.

About SME: The Society of Manufacturing Engineers is the world's leading professional society supporting manufacturing education. Through its member programmes, publications, expositions and professional development resources, SME promotes an increased awareness of manufacturing engineering and helps keep manufacturing professionals up to date on leading trends and technologies. Headquartered in Michigan, SME influences more than half a million manufacturing practitioners and executives annually. The Society, which celebrates its 75th anniversary in 2007, has members in more than 70 countries and is supported by a network of hundreds of technical communities and chapters worldwide.

About Battelle: Battelle is the world's largest non-profit independent research and development organization, with 20,000 employees in more than 120 locations worldwide, including five national laboratories Battelle manages or co-manages for the US Department of Energy. Headquartered in Columbus, Ohio, Battelle conducts $3.8 billion in R&D annually through contract research, laboratory management and technology commercialization. Battelle provides innovative solutions to some of the world's most important problems including global climate change, sustainable energy technologies, high performance materials, next generation healthcare diagnostics and therapeutics and advanced security solutions for people, infrastructure and the nation. Battelle has a long history of developing successful commercial products in collaboration with its clients, ranging from products to fight diabetes, cancer and heart disease to the development of the office copier machine (Xerox). As a non-profit charitable trust with an eye toward the future, Battelle actively supports and promotes science and math education.

Foresight Nanotechnology Challenges

While nanotechnology has made great advances in the last two decades, it has yet to fulfill its ultimate potential. Foresight is dedicated to fostering nanotechnologies that can make a significant contribution to solving critical challenges which humanity faces- Foresight Nanotechnology Challenges. These challenges are:

Providing renewable clean energy

Supplying clean water globally

Improving health and longevity

Healing and preserving the environment

Making information technology available to all

Enabling space development

Legal Implications

The Oklahoma land rush of 1889 and similar land runs were events in which previously restricted land of the United States was opened for homesteading on a first come, first claim basis. A number of the individuals who participated in the run entered early and hid out until the legal time of entry to lay quick claim to some of the most choice homesteads. These people came to be identified as "sooners" and their actions led to hundreds of legal contests. Nowadays, universities and companies, from start-ups to multinational conglomerates, are rushing to aggressively stake out their turf in the nanotechnology patent area. The "sooners" in this nanotechnology patent land rush may be the ones who were issued what some say are "unduly broad" patents early on, in the hope of getting a windfall of nanotechnology intellectual property (IP) rights.

While not violating the rules intentionally, upon re-examination some broad patents might not hold up to the United States Patent Act's requirement for full and complete disclosure. This uncertainty provides fertile grounds for possible litigation over nanotechnology patent claims based on broad and unprecise definitions and descriptions. Efforts are underway to address some of these problems.

Start of the 1889 Oklahoma Land Rush

Reflecting the growing concerns over this issue, we have covered nanotechnology patent rights in a number of Nanowerk Spotlight articles over the past year, from a general overview on IP rights (Nanotechnology and intellectual property issues), specific issues with patents in the carbon nanotube area (Growing nanotechnology problems: navigating the patent labyrinth) to the sheer numbers of nanotechnology patents issued (The patent land grab in nanotechnology continues unabated, creating problems down the road).

"As scientists sort out and document the results of their research, corporate entities continue to seek and carve out far-reaching patent rights in what is now a full scale patent 'land grab'" Dr. Raj Bawa tells Nanowerk. "As this trend unfolds, uncertainty is growing amongst researchers, developers, policy-makers and investors regarding who really owns what particular swath of technology in the rapidly-expanding body of nanotechnology intellectual property. Some fear that the far-reaching patent rights provided by early nanotechnology patents clearly overlap."

Bawa is a registered patent agent and holds a faculty position at Rensselaer Polytechnic Institute in Troy, New York where he also serves as advisor to the Office of Technology Commercialization. Together with several colleagues from the legal field he co-authored a recent article titled "Broad Claiming in Nanotechnology Patents: Is Litigation Inevitable?".

Bawa points out that some commentators, ranging from university experts to government agencies, blame this trend of uncertainty and patent overlaps on problems at the US Patent & Trademark Office (USPTO), including a delay in implementing nanotechnology training for examiners. Further contributing to this uncertainty is the granting of patents of questionable validity and scope, as well as a growing backlog of unexamined patent applications and increasingly lengthy periods for patent pendency.

"Add to this backdrop the limited number of judicial opinions on patents involving nanotechnology and a lack of standardized terminology and you have a patent landscape that is almost impossible to navigate in certain nanotechnology sectors like for instance carbon nanotubes" says Bawa.

The USPTO has started to address issues from its end by adding a new cross-reference digest to its Class 977 Nanotechnology section. As part of its continuing effort to improve the ability to search and examine nanotechnology-related patent documents, the agency has created over 250 cross-reference art collection subclasses for nanotechnology.

A major issue in potential legal conflicts over patent claim coverage are the definitions used in the patent applications. In the Wild West of nanotechnology terminology there is considerable confusion over what is what and there isn't even a clear and generally used definition for the term nanotechnology itself. The same is true for such terms as nanoparticles, nanostructures or nanomaterials, just to name a few. Usage of such broad terms and ambiguous descriptions of even more specific technical terms such as carbon nanotubes or quantum dots in patent applications are a sure recipe for conflicting terminology and a dispute over what was meant and intended.

Bawa explains that, even if a patent has been issued already, a claimant could try to dispute it and get it invalidated if the original patent application violates the United States Patent Act at 35 U.S.C. §112. This paragraph contains three distinct requirements for patentability: "(1) a complete written description—setting forth the full and complete details of making and using the invention; (2) enablement—the specification must teach one skilled in the art how to make and use the full scope of the invention without undue experimentation and (3) best mode—of carrying out the claimed invention known at the time of filing for a patent needs to be set forth." A claimant could try to get the original claim re-interpreted by a court and argue that the use of broad or unspecific terminology in early, broad patents might violate §112.

Owners of nanotechnology patents that lack a clear definition of claim terms may have to live with the uncertainty as to whether someone will dispute their claim at some point down the road. As a general strategy to avoid downstream problems, Bawa and his colleagues suggest that applicants file patent applications that are not unnecessarily broad in scope.

"Patent applicants must resist the urge to claim overly broadly in the hope of getting a windfall of nanotechnology IP rights" says Bawa. "Such attempts may be responsible in part for creating uncertainty as to who owns what across the nanotechnology landscape."

Efforts to develop a common terminology for nanotechnology (such as ASTM's Terminology for Nanotechnology Standard), improved classification systems from patent offices and self-constraint from patent applicants will go a long way in reducing the existing uncertainties in the nanotechnology patent field.

Work with Nanoparticles

Chemical engineers from the University at Buffalo [profile] have collaborated with scientists from other institutions to solve a critical bottleneck in the transport and capture of virus nanoparticles, making possible a device that could rapidly sample and detect infectious biological agents, such as viruses.

"This advance may pave the way for an 'on-the-spot' virus detector, which would be immensely helpful, especially in military and public-health applications," said Paschalis Alexandridis, Ph.D., professor in the Department of Chemical and Biological Engineering in the UB School of Engineering and Applied Sciences and co-author on the research.

A paper describing the results was published in the March issue of Langmuir (vol. 23, p. 3840).

The rapid detection of viruses in biological samples is of increasing interest, particularly with the recent emergence of new viruses, including SARS, West Nile virus and avian flu virus.

But because viral particles are present at such low concentrations in biological samples, such as blood, a device that can quickly and easily detect them has remained elusive.

Typical procedures involve using passive diffusion to get the viral particles to bind to an antibody, a slow process that is not feasible for many applications, such as on the battlefield, where quick results are critical.

Scientists at the University of Wisconsin at Madison [profile] led by Nicholas L. Abbott, Ph.D., a co-author on the paper and John T. and Magdalen L. Sobota Professor of Chemical and Biological Engineering, previously had demonstrated that liquid crystals can amplify signals from low concentrations of viral particles, quickly indicating whether or not a virus is present on a surface.

"The bottleneck was how to transport and capture enough suspected viral particles onto a surface in a timely fashion so that they could be detected," said Alexandridis. "During the acute phase of an infection, the virus is at a very low concentration and relying on passive diffusion to deposit the viral particles onto the detection surface can be time-consuming."

The researchers wanted to speed up the collection of viral particles—in this case, of vesicular stomatitis virus, a common animal virus—at the right place on a substrate, while also doing it in media at physiological ionic strength.

The UB researchers used their expertise in a technique called directed assembly, in which they design external electrical and fluid flow fields in order to "drive" nanoparticles to specific locations and in specific concentrations on a substrate.

"This paper shows that by using electrodes separated by just a few micrometers together with electrothermally induced fluid flow, we can accelerate the transport of viral particles from aqueous suspensions with physiological ionic strength to specific points on a surface, allowing them to reach local concentrations high enough to allow subsequent rapid detection," Alexandridis explained.

"We achieved this not by accident, but by design," he continued. "We hypothesized that the application of these external fields would cause the nanoparticles to act in a certain way. We designed electrodes to generate the required forces for the system of interest and then put our design to the test."

In the research, the UB engineers used directed assembly to tailor dielectrophoretic forces, which act through a nonuniform electric field, overcoming an obstacle that occurs whenever nanoparticles are involved.

"When you work with microscopic objects dispersed in a liquid, gravity is very important," explained Alexandridis. "But at the nanoscale, gravity doesn't matter. So when you are trying to manipulate matter at the nanoscale, electrical fields and fluid fields may work best. By using directed assembly, we can tailor the forces acting on the nanoparticles. The ability to use several forces acting in tandem becomes important."

Electrical fields in particular, he said, are advantageous because by designing the electrodes in a certain way, engineers can control directionality and intensity of electrical forces acting on nanoparticles.

Alexandridis conducted the research with Aristides Docoslis, Ph.D., formerly a doctoral candidate and postdoctoral associate in Alexandridis' lab at UB and now the Canada Research Chair of Colloids and Surface Engineering at Queen's University at Kingston.

Additional co-authors are Luis A. Tercero Espinoza, Ph.D., of the Department of Chemical and Biological Engineering, University of Wisconsin at Madison; Bingbing Zhang, Ph.D., of Queen's University at Kingston and Li-Lin Cheng, Ph.D. and Barbara A. Israel, Ph.D., of the Veterinary School of the University of Wisconsin at Madison.

The research was funded by a National Science Foundation [profile] grant in Nanoscale Exploratory Research and by the Gerald A. Sterbutzal Research Fund, administered by the UB School of Engineering and Applied Sciences.

The University at Buffalo is a premier research-intensive public university, the largest and most comprehensive campus in the State University of New York.

Tiny Wires Power Nano Revolution

Professor Zhong Lin Wang of the Georgia Institute of Technology in the US developed the tiny generator to harvest mechanical energy from its surroundings.

"This is a major step toward a portable, adaptable and cost-effective technology for powering nanoscale devices," said Zhong Lin Wang, Regents' Professor in the School of Materials Science and Engineering. "There has been a lot of interest in making nanodevices, but we have tended not to think about how to power them. Our nanogenerator allows us to harvest or recycle energy from many sources to power these devices."

The prototype is made up of wires that are free to flex just slightly. The nanogenerator takes advantage of the unique coupled piezoelectric and semiconducting properties of zinc oxide nanostructures, which produce small electrical charges when they are flexed. By capturing the output of large numbers of nanowires in motion, the nanogenerator produces a miniscule direct current output.

Wang and his group members expect that their nanogenerator could produce as much as 4 watts per cubic centimeter—based on a calculation for a single nanowire. That would be enough to power a broad range of nanometer-scale defense, environmental and biomedical applications, including biosensors implanted in the body, environmental monitors and even nanoscale robots.

However, before they can do that, the team needs to optimise the development process.

"We need to be able to better control the growth, density and uniformity of the wires," Wang said. "We believe we can make as many as millions or even billions of nanowires produce current simultaneously. That will allow us to optimize operation of the nanogenerator."

Nanotech Soon to Curb Diesel Abuse in Sabah

Kota Kinabalu: Nanotechnology will soon be used to tag subsidised diesel to combat diesel smuggling in Sabah.

Deputy Minister for Innovations, Science and Technology, Datuk Kong Cho Ha, who revealed this Thursday, said it would also ensure that there is no misuse by certain sectors due to the low-price of the commodity compared to unsubsidised diesel.

Presently, nano-technology, which has to do with the science of building devices like electronic circuits from single atoms or molecules, is only employed in Kuala Lumpur.

To a question, Kong, who was on a visit to the Chemistry Department's branch office here, said there were however no plans to set up a deoxyribonucleic acid (DNA) lab in the State for the time being, citing high cost as a main factor.

"The Ministry, however, will consider the matter since DNA has been proven to be very important and reliable in solving many cases," he said, citing one example of a case solved, thanks to DNA taken from a toothpick. The two DNA labs at present are in Petaling Jaya, Selangor and Kuching, Sarawak. Another will be built in Penang to cater to the northern regions of Peninsular Malaysia, he said.

He also announced that a second branch of the Chemistry Department will soon be set up in the State to provide better assistance to enforcement agencies here.

"Sabah is developing rapidly and the Ministry sees the need for a second branch in the State," he said.

The Ministry is in the process of studying a location in Tawau and the matter will be finalised soon, he added.

Presently, there is only one Chemistry Department branch operating here with 62 staff comprising 39 technical and 23 administration personnel. He said he was satisfied with what the department has done so far and that he was open to suggestions on ways to improve the department's service.

The department's Sabah Director, Wong Kok Fah, who took Kong on a guided tour of the facilities, said the department's objective was to assist government agencies in dealing with the issues of security, health, environment, consumer affairs, tariff classifications and also to uphold justice.

The scientific services offered by the department were forensic analysis (encompassing narcotics, toxicology, serology and criminalistic), food, water as well as industry analysis.

Wong pointed out the police were the department's main customers, followed by the Health Ministry, Customs and Environment departments, Ministry for Domestic Trade and Consumer Affairs as well as the Road Transport Department.

He refuted claims that the department was responsible for the many unresolved drug-related cases in the courts, saying the department adhered to its Client Charter on returning samples in the stipulated time.

In this case, the department issues the analysis and reports back to the relevant agencies in one-month, he said, adding Syabu (methamphetamine) accounted for most of the drugs that were sampled and analysed.

Nanotube Needles Deliver Quantum Dots

Multiwalled carbon nanotubes can be used as nanoinjectors to deliver quantum dots inside cells, a technique that could enhance studies of cellular processes (Proc. Natl. Acad. Sci. USA, DOI: 10.1073/pnas.0700567104).

Chemistry professor Carolyn R. Bertozzi, chemistry graduate student Xing Chen, physics professor Alex Zettl and physics postdoc Andras Kis at the University of California, Berkeley [profile] and Lawrence Berkeley National Laboratory [profile] created the devices by first connecting a single nanotube to an atomic force microscope tip. A quantum dot is attached to the nanotube by a linker that has a pyrene moiety at one end, a biotin moiety at the other end and a disulfide bond in the middle. The pyrene adsorbs to the nanotube, while the biotin binds to streptavidin on the surface of the quantum dot.

As the nanotube needle pokes through a cell membrane, the reducing environment of the cytosol cleaves the disulfide bond, which releases the quantum dot inside the cell. The researchers used single-particle tracking methods to follow quantum dots injected into human cells. This delivery method doesn't require the use of carrier solvents and appears not to physically harm the cells, they note.

Sr. Engineer—Component Development

Job Description: Work within project group to implement key hardware and process inventions in form suitable for advanced prototyping and manufacturing. Will play critical role in development of essential support components needed to enable fuel cell technology performance. Design and test chemical processing assemblies for fuel cell related functions. Accumulate and present supporting data and documentation to executive level for project gateway review.

Job Requirements: MS chemistry, chemical engineering or mechanical engineering. 3+ years non-academic (including national lab unless for company-sponsored project) experience in fuel cell, medical device, analytical or biochemical arenas. Prefer experience in microfluidics. Must be able to show experience of a leading role in development of manufacturable chemical handling assemblies from the research stage. Experience in electrochemical or electroanalytical techniques is preferred.

FEI Company: Staff Scientist 886-HBO

Job Description: The Company

FEI Company is a leading supplier of Tools for Nanotech enabling research, development and manufacture of products at the Nanoscale. Our range of industry-leading focused ion and electron beam hardware and software products are delivered to three primary Nanotechnology markets: NanoElectronics, NanoResearch and NanoBiology. With R&D centres in North America and Europe and sales and service operations in more the 40 countries around the world, FEI is bringing the Nanoscale within the grasp of leading researchers and manufacturers and helping to turn some of the biggest ideas of the this century into reality.

FEI Company is a publicly held company with over 1700 employees worldwide. At FEI Company we are committed to our employees and to diversity and innovation in the workplace.

The Position

Charged particle optic electron and ion guns, focusing columns and high-brightness sources are at the heart of FEI's Scanning Electron Microscope (SEM), Focused Ion Beam (FIB) and DualBeam tools serving the Nanotechnology market. FEI seeks a staff level scientist to be responsible for the initiation, design, development, execution and implementation of a wide range of electron gun, ion column and field emission source development projects.

Position Description:

- Support a variety of electron gun, ion column and high-brightness source research and development projects through:
- Theoretical analysis of electromagnetic fields, high voltage design, magnetic shielding, mechanical design, vacuum analysis, thermal analysis and material selection.
- Simulation and design of charged particle optic columns for focused electron and ion beam systems.
- Design and execution of experiments to prove column performance and develop new or improved field emission sources.
- Assist in optimizing system-level performance of Beam Technology developed products.
- Communication and documentation of test results and milestones via presentations and reports.
- Apply sound scientific principles and concepts, via both analysis and experimentation, to potential inventions and products.
- Maintain technical currency in a broad range of state of the art principles and theories relevant to charged particle optics.
- Assist management in the development of technology and strategic roadmaps for FIB, Electron Gun, Electron & Ion Source and Beam Chemistry technologies.

- o Contribute regularly to scientific literature and conferences.
- o Develop patent applications for work where appropriate.
- o Manage small- to medium-sized electron gun, ion column and source development projects.
- o Travel periodically as necessary to fulfill job requirements.

Job Requirements: The Requirements

Minimum Education, Experience and Qualifications:

- o PhD, or equivalent experience, in Physics, Chemistry, or related field.
- o 10+ years laboratory or related experience.
- o Expertise in charged particle optics and/or high-brightness source technology.
- o Expertise in charged particle optics theory, design and implementation.
- o Excellent verbal and written communication skills.
- o Experience in leading project teams and/or managing small groups.
- o Experience in designing and conducting experiments.
- o Experience troubleshooting complex instrumentation.
- o Experience operating SEMs, FIBs or similar systems.
- o Experience with Ultra-High Vacuum (UHV) and High Voltage (HV) principles and practices.
- o Proficient in use of measuring equipment (oscilloscope, multimeter, HV probes, etc.)
- o Aptitude for creative problem solving in a developmental lab environment.
- o Ability to work in a team environment.
- o Ability and willingness to travel internationally, eligible to obtain passport.

Additional Desirable Qualifications

- o Expertise in high-brightness electron and ion source design and/or theory.
- o Experience with charged particle optics modelling software programmes.

- o Familiarity with materials issues relating to high vacuum, high voltage and high precision mechanical assemblies.
- o Ability to programme in Visual Basic, Lab View and/or C/C++.

Charles Stark Draper Laboratory

Descriptions & Responsibilities: Draper currently has major MEMS efforts in inertial sensing, RF devices for communications, nanotechnology and microfluidics. We are seeking a senior level staff member with direct experience in the specification, design and test of MEMS devices for inertial sensing applications. We offer an atmosphere which fosters innovation, room for individual contributions within a team environment and an emphasis on professional and intellectual development.

Job Requirements: Education & Experience Required: We seek individuals with an M.S. or preferably Ph.D. in electrical engineering, chemistry, biophysics, chemical engineering, mechanical engineering, or a related discipline. The candidate should have practical laboratory experience in microfabrication and test. The individual should have demonstrated an ability to work autonomously, write proposals, design experiments, analyze data and relate research efforts to current scientific literature. Candidates should have a minimum of 6 years of post-PhD experience or 9 years of post-MS experience.

Extra Information: Applicants selected will be subject to a government security investigation and must meet eligibility requirements for access to classified information.

Senior Synthetic Chemist, Semiconductor Realm

Job Description: PRISM Group is actively seeking an experienced, senior-level chemist with a diverse background that includes complex organic, organometallic and inorganic synthesis, materials and surface analysis, electrochemical characterization and at least 1-2 years of supervisory experience. Responsibilities include (but are not limited to) synthesis of organometallic molecules. The successful candidate must be willing to work in the laboratory as an individual contributor, in team settings and supervise small groups of chemists and engineers of various disciplines and degree levels. Demonstrated creativity, leadership and good communication skills are musts. A Ph.D. in Chemistry and at least 2+ years of experience working in industrial labs are required. Working knowledge of surface chemistry, solid phase synthesis, organometallics are preferred. US citizenship and problem solving analytical skills are important qualifications.

Job Requirements: PRISM Group is actively seeking an experienced, senior-level chemist with a diverse background that includes complex organic, organometallic and inorganic synthesis, materials and surface analysis, electrochemical characterization and at least 1-2 years of supervisory experience. Responsibilities include (but are not limited to) synthesis of organometallic molecules. The successful candidate must be willing to work in the laboratory as an individual contributor, in team settings and supervise small groups of chemists and engineers of various disciplines and degree levels. Demonstrated creativity, leadership and good communication skills are musts. A Ph.D. in Chemistry and at least 2+ years of experience working in industrial labs are required. Working knowledge of surface chemistry, solid phase synthesis, organometallics are preferred. US citizenship and problem solving analytical skills are important qualifications.

Post Doctoral Associate in MEMS Sensors

Job Description: UTA-NanoFab has an immediate opening for a full-time postdoctoral researcher in the area of force and acceleration microsensors. The initial appointment is for 8 months, with the possibility of extension for 1-1A additional years depending on the candidate's performance and the availability of funds. Very competitive salary and benefits. Applications will be accepted until the position is filled.

Job Requirements: Candidates with strong theoretical and experimental background/experience in one or more of the following areas are sought: force and pressure microsensors, piezoresistors and accelerometers. Hand-on experience in MEMS fabrication is a plus. To be considered, candidates must have published in at least one of the aforementioned areas, must have a PhD degree in EE/ME by the time of appointment and must have strong communication skills in English.

Microfabrication and Characterization

Job Description: A postdoctoral position in the area of microfluidic devices measurements is currently available at CU-Boulder, in the Department of Aerospace Engineering Sciences. The position is available immediately and will be filled as soon as a suitable candidate is identified.

Job Requirements: We are seeking candidates who can combine standard microfabrication skills, such as optical lithography and different deposition methods, to make structures for testing the microfluidic devices using Electrowetting on Dielectric (EWOD) technology. The project is focused on using EWOD, DEP and Continious Electrowetting for various applications. Experience is Atomic Layer Deposition is a plus. Strong hand-on experience is required. Background with heat transfer and fluidic mechanics is required. Experience with programming with LabView, Matlab, or C++ and electrical circuit is a plus. Strong English communication skills are also very important. Candidates must have completed a PhD in Mechanical Engineering, Physics, Electrical Engineering, or a related field by the time of appointment.

EG&G Technical Services, Inc.: Scientist

Job Description: Job responsibilities will include: developing test procedures and conducting research experiments in the area of aersol research relevant to personal protective equipment and respiratory protection; assembling/operating/maintaining test equipment, including aerosol test systems, particle generators, particle counters, filter testers an other related aerosol equipment; Maintaining lab notebooks; Preparing and characterizing aerosol particles that will be used in research experiments and independently collecting experimental data; Preparing required documentation and assists investigators in conducting research projects. Responsibilities will also include giving presentations at scientific meetings and writing papers for publication in national and international journals.

Job Requirements: BSci. in Engineering, Physics, Mathematics or related scientific field (MS or PhD in aerosol technology preferred) and 10 years experience or equivalent. Candidate must have working knowledge of aerosol science and technology with lab experience and thorough understanding of the principles and fuction of aerosol generation systems and measurement techniques with the ability to independently assemble and troubleshoot aerosol test systems (*i.e.*, TSI aerosol generators, fit test equipment, TSI 3160) and develop research protocols and projects in respiratory protection, filtration theory, or aerosol science. Candidate must also be able to lift and carry equipment weighing 30 pounds for short distances. Position requires the ability to pass and maintain a security clearance.

Sr. Process Engineer (Nanotechnology)

Descriptions & Responsibilities: Draper currently has major MEMS efforts in inertial sensing, RF devices for communications and microfluidics and nanotechnology/ nanofabrication programmes in sub-wavelength optics and plasmonics, nanopatterened polymers for control of surface properties and integration of carbon nanotubes with MEMS. We are seeking a senior level staff member with direct experience in nanofabrication techniques and device design. We offer an atmosphere which fosters innovation, room for individual contributions within a team environment and an emphasis on professional and intellectual development.

Job Requirements: Education & Experience Required: We seek individuals with an M.S. or preferably Ph.D. in electrical engineering, chemistry, biophysics, chemical engineering, mechanical engineering, or a related discipline. The candidate should have practical laboratory experience in micro-and nanofabrication and test.

The individual should have demonstrated an ability to work autonomously, write proposals, design experiments, analyze data and relate research efforts to current scientific literature. Candidates who have a minimum of 6 years of post-Ph.D. experience or 9 years of post-MS experience are strongly encouraged to apply.

Senior Device Physics Engineer

Job Description: PRISM Group is actively seeking Device Physics Engineers with solid CMOS experience. Title: Principal Device Engineer Requirements: Ivy league or similar education required—Ph.D. in Electrical Engineering with expertise in experimental semiconductor device physics, device design and device modelling—5 years industry or academic experience in CMOS process integration / device reliability is desirable. Knowledge of organic chemistry and materials will be an asset—Must be a strong hands-on individual contributor and a self-starter.

Must have sound analytical/systematic thinking—Expected to work with a cross-functional team of chemists, physicists, material scientists and electrical engineers to optimize memory cells. Good communication skills are required.—Expected to define, conduct and develop experiments to modify and improve device structure—The position could be based in the Denver, CO area or the Bay area. Travel between the two would be required. Successful people will have Flash EProm, or general memory industry experience including folks from the cellular realm.

Postdoctoral Fellow

Job Description: Full responsibility in a new well-funded project on the synthesis and characterization of novel nanostructured materials prepared by self-assembly. Pore-engineering in zeolites. Applications in catalysis and adsorption. One-year position extendible to two years. Salary based on experience and qualifications.

Job Requirements: The ideal candidate will have experience in the preparation and characterization of porous materials, a solid Chemistry/Chemical Engineering background and an good publication track record.

Bibliography

Albarran, Alan B. and Gregory G. Pitts: *The Radio Broadcasting Industry*, Boston, Allyn and Bacon, 2001.

Albrecht, Ulrich: *A Short Research Guide on Arms and Armed Forces*, New York, Facts on File, 1980.

Alexander, James: *Internships in Communications, Ames*, Iowa State University Press, 1995.

Allman, Paul: *Careers in Video and Digital Video*, New York: Rosen, 2001.

Anders, Peter: *Envisioning Cyberspace*, New York, McGraw-Hill, 1999.

Arkin, William M.: *Research Guide to Current Military and Strategic Affairs*, Washington, Institute for Policy Studies, 1981.

Baehr, H. and Gary A.: *Turning In On: A Reader in Women and Media*, London, Arnold, 1996.

Ball, Victoria Kloss: *Opportunities in Interior Design and Decorating Careers*, Chicago, VGM Career Horizons, 2002.

Barnouw, Erik: *A Tower in Babel: the History of Broadcasting in the United States to 1933*, New York, Oxford University Press, 1966.

Baucom, Alfred H.: *Hospitality Design for the Graying Generation: Meeting the Needs of a Growing Market*, New York, Wiley, 1996.

Baxter, John: *Australian Cinema*, Sydney, Pacific Books, 1970.

Beaman, Jim: *Interviewing for Radio*, New York, Routledge, 2000.

Berkowitz, Ira: *Vault Career Guide to Advertising*, New York, Vault, 2004.

Bertrand, Ina and Diane Collins: *Government and Film in Australia*, Sydney, Currency Press, 1981.

Bliss, Edward and James L. Hoyt: *Writing News for Broadcast*, New York, Columbia University Press, 1994.

Boella, Michael John: *Principles of Hospitality Law*, London, Cassell, 1999.

Bone, Jan, & Ana Fernandez: *Opportunities in Film Careers*, New York, VGM Career Books, 2004.

Bowers, C. A.: *The Cultural Dimension of Educational Computing: Understanding the Non-neutrality of Technology*, New York, Teachers' College Press; 1988.

Boyd, Andrew: *Broadcast Journalism: Techniques of Radio and TV News*, Woburn, Focal Press, 2001.

Brown, Douglas Robert: *The Restaurant Managers Handbook: How to Set Up, Operate, and Manage a Financially Successful Restaurant*. Lauderhill, Atlantic Pub., 1991.

Brown, James and Quaal Ward: *Radio-Television-Cable Management*, New York, McGraw Hill, 1998.

Bryant, Jennings: *Television and the American Family*, Hillsdale, Lawrence Erlbaum, 1990.

Buckingham, David: *Young People and the Media*, New York, Manchester University Press, 1993.

Buckner, Robert W.: *Site Selection: New Advancements in Methods and Technology*, New York, Lebhar-Friedman Books, 1998.

Caldwell, J. T.: *Electronic Media and Technoculture*, New Brusnwick, Rutgers University Press, 2000.

Careers in Focus: *Photography*, New York, Ferguson, 2004.

Carmouche, Rita: *Behavioural Studies in Hospitality Management*, New York, Chapman & Hall, 1995.

Carroll, Victoria: *Writing News for Television*, Ames, Iowa State University Press, 1997.

Cassiday, Doris, and Bruce Cassiday: *Careers in the Beauty Industry*, New York, Franklin Watts, 1978.

Clark, Mona: *Interpersonal Skills for Hospitality Managers*, London, Chapman Hill, 1995.

Clark, Roy Peter: *A Journalist Teaches Young Writers*, Portsmouth, Heinemann, 1987.

Cockburn, C. and Ormrod, S.: *Gender and Technology in the Making*, London, Sage, 1993.

Cohler, David Keith: *Broadcast Newswriting*, Englewood Cliffs, Prentice Hall, 1990.

Cracknell, H. L.: *Escoffier: The Complete Guide to the Art of Modern Cookery*, New York, John Wiley, 1979.

Craig, James: *Graphic Design Career Guide*, New York, Watson-Guptill, 1992.

Crawford, Tad: *Starting Your Career As a Freelance Photographer*, New York, Allworth Press, 2003.

Crouse, Chuck: *Reporting for Radio*, Chicago, Bonus Books, 1998.

Davidson, M. J. and Cooper, C. L.: *Women and Information Technology*, New York, John Wiley & Sons, 1987.

Delaney, Chuck: *Photography Your Way*, New York, Allworth Press, 2005.

Dennis, Everette E. and Edward C. Pease: *Children and the Media*, New Brunswick, Transaction Publishers, 1996.

Diamond, S. Taylor's Way: *Women, Cultures and Technology*, New York: Routledge; 1997.

Donald, Ralph and Thomas Spann: *Fundamentals of Television Production*, Ames, Iowa State University Press, 2000.

Dornenburg, Andrew: *Culinary Artistry*, New York, John Wiley & Sons, 1996.

Douglas, George H.: *The Early Days of Radio Broadcasting*, Jefferson, NC: McFarland, 1987.

Easlea, B.: *Fathering the Unthinkable: Masculinity, Scientists and the Nuclear Arms Race*, London, Pluto Press, 1982.

Ellis, Elmo: *Opportunities in Broadcasting Careers*, Chicago, VGM Career Books, 2005.

Faulkner, W. and Arnold, E.: *Smothered by Invention: Technology in Women's Lives*, London, Pluto Press, 1985.

Fensch, Thomas: *Television News Anchors*, Jefferson, McFarland, 1993.

Field, Shelly: *Career Opportunities in Advertising and Public Relations*, New York, Ferguson, 2006.

Foster, Timothy R. V.: *How to Succeed As an Independent Consultant*, London: Kogan, Page, 2002.

Fox Keller, E.: *Reflections on Gender and Science*, New Haven, Yale University Press, 1985.

Gardner, Garth: *Careers in Computer Graphics and Animation*, Fairfax, GGC, 2001.

Gibbs, Jenny: *Interior Design*, New York, H. N. Abrams, 2005.

Go, Frank M.: *Human Resource Management in the Hospitality Industry*, New York, Wiley, 1996.

Goldfarb, Roz: *Careers by Design*, New York, Allworth Press, 2001.

Gordon, Sandra R.: *Great Jobs for Film Majors*, Chicago, VGM Career Books, 2004.

Graham, Duane: *The Remarkable Story of Otto Graham*, Wayne, Immortal Investments Publishing, 2004.

Green, E. and Adam, A. Virtual Gender: *Technology, Consumption, and Identity*, New York, Routledge, 2001.

Greenman, Robert: *The Adviser's Companion*, New York, Columbia Scholastic Press Association, 1991.

Griffiths, D.: *The Exclusion of Women from Technology*, London, Pluto Press, 1985.

Hall, Ken G.: *Australian Film: the Inside Story*, Sydney, Summit Books, 1980.

Hewitt, John: *Airwords: Writing for Broadcast News, Mountain View*, Oxford, Mayfield Pub. Company, 2002.

Higham, Robin: *A Guide to the Sources of United States Military History*, Hamden, Archon Books, 1975.

Hilliard, Robert L. and Michael C. Keith: *The Broadcast Century and Beyond*, Boston: Focal Press, 2001.

Hinkin, Timothy R.: *Cases in Hospitality Management: A Critical Incident Approach*, New York, Wiley, 1995.

Holloway, J. C. : *Towards a Core Curriculum for Tourism: A Discussion Paper*, London, The National Liaison Group for Higher Education in Tourism, 1995.

Horstede, G.: *Cultures and Organizations: Software of the Mind*, London, McGraw-Hill Book Company, 1991.

Hungerland, Buff: *Marketing Your Creative Portfolio*, Upper Saddle River, Prentice Hall, 2003

Jain, S. C.: *Women and Technology*, Jaipur, Rawat Publications, 1985.

Jefferson, Michael: *Breaking into Graphic Design*, New York, Allworth Press, 2005.

Jensen, Clayne R.: *Opportunities in Recreation and Leisure Careers*, Lincolnwood, VGM Career Horizons, 2000.

Jones, Sue Jenklyn: *Fashion Design*, New York, Watson-Guptill, 2002.

Kieffer, John: *The Photographer's Assistant*, New York, Allworth, 2001.

Kirkham, P.: *The Gendered Object*, Manchester, Manchester University Press, 1996.

Kissane, Sharon F.: *Career Success for People with Physical Disabilities*, Lincolnwood, VGM Career Horizons, 1997.

Kring, Robin: *Party Creations: Book of Theme Event Design*, Denver, Clear Creek Pub., 1993.

Kurzweil, R.: *The Age of Spiritual Machines*, New York, Viking, 1999.

Lavery, P.: *Tourism Marketing and Management Handbook*, New York, Prentice Hall, 1989.

Lewis, Robert C.: *Cases in Hospitality Marketing and Management*, New York, John Wiley, 1997.

MacCloskey, Monro: *How to Qualify for the Service Academies*, New York, Richards Rosen Press, 1964.

Maclay, Edgar Stanton: *A History of the United States Navy 1771-1893*, New York, Appleton Co., 1894.

Mauro, Lucia: *Careers for Fashion Plates and Other Trendsetters*, Chicago, VGM Career Horizons, 2003.

McGuire-Lytle, Erin: *Careers in Graphic Arts and Computer Graphics*, New York, Rosen, 1999.

Miller, Samuel D.: *An Aerospace Bibliography*, Washington. Office of Air Force History, 1978.

Mitchell, Leslie Scott: *Freelancing for Television and Radio*, London; Routledge, 2005.

Mogel, Leonard: *Creating Your Career in Communications and Entertainment*, Sewickley, GATF Press, 1998.

Noronha, Shonan F. R.: *Opportunities in Television and Video Careers*, Chicago: VGM Career Books, 2003.

Orlik, Peter B.: *Career Perspectives in Electronic Media*, Ames, Blackwell Pub., 2004.

Packard, Sidney: *The Fashion Business: Dynamics and Careers*, New York, Holt, Rinehart, and Winston, 1983.

Parks, Peggy J.: *The News Media*, San Diego, Lucent Books, 2002.

Paullin, C.O.: *The Navy of the American Revolution*, Cleveland, The Burrows Brothers Co., 1906.

Perna, Rita: *Fashion Forecasting*, New York, Fairchild Publications, 1987.

Rachlis, Eugene: *The Story of the U.S. Coast Guard*, New York, Random House, 1961.

Reis, Ronald A.: *Careers in Art and Graphic Design*, Hauppage, Barrons, 2001.

Rocha, Tony L.: *Careers in Magazine Publishing*, New York, Rosen, 2001.

Rubinstein, Donna: *The Modeling Life*, New York, Berkeley Publishing Group, 1998.

Satterthwaite, Frank: *The Career Portfolio Workbook*, New York, McGraw-Hill, 2003.

Seguin, James: *Media Career Guide*, Boston, St. Martin's, 2006.

Sones, Melissa: *Getting into Fashion: A Career Guide*, New York, Ballantine, 1984.

Vogt, Peter: *Career Opportunities in the Fashion Industry*, New York, Facts on File, 2002.

Wilson, Wayne: *Careers in Publishing and Communications*, Bear, Mitchell Lane, 2002.

Yager, Fred, & Jan Yager: *Career Opportunities in the Film Industry*, New York, Ferguson, 2003.

Zahler, Karen Gantz.: *Superchefs: Signature Recipes From America's New Royalty*, New York, John Wiley & Sons, 1996.

Index

D

M

N

O

T

U

W